Robust Statistical Methods with R

Second Edition

T0227990

Robust Statistical Methods with R

Second Edition

Jana Jurečková
Jan Picek
Martin Schindler

CRC Press
Taylor & Francis Group
Boca Raton London New York

CRC Press is an imprint of the
Taylor & Francis Group, an **informa** business

A CHAPMAN & HALL BOOK

CRC Press
Taylor & Francis Group
6000 Broken Sound Parkway NW, Suite 300
Boca Raton, FL 33487-2742

First issued in paperback 2021

ISBN-13: 978-1-03-209260-7 (pbk)
ISBN-13: 978-1-138-03536-2 (hbk)

Library of Congress Cataloging-in-Publication Data

Names: Jurečková, Jana, 1940- author. | Picek, Jan, 1965- author. |
Schindler, Martin (Mathematician), author.
Title: Robust statistical methods with R / Jana Jurečková, Jan Picek,
Martin Schindler.
Description: Second edition. | Boca Raton, Florida : CRC Press, [2019] |
Includes bibliographical references and index.
Identifiers: LCCN 2019006741| ISBN 9781138035362 (hardback : alk.
paper) | ISBN 9781315268125 (e-book : alk. paper)
Subjects: LCSH: Robust statistics. | Mathematical statistics. | R (Computer
program language)
Classification: LCC QA276 .J868 2019 | DDC 519.5--dc23
LC record available at https://lccn.loc.gov/2019006741

Visit the Taylor & Francis Web site at
http://www.taylorandfrancis.com

and the CRC Press Web site at
http://www.crcpress.com

Contents

Preface

The present text on robustness is a follow-up of the earlier textbook of the first two authors (J. Jurečková and J. Picek (2006): *Robust Statistical Methods with R*, Chapman & Hall/CRC. The theoretical background of the robust statistical procedures and their initial applications developed mainly during the 1960s and 1970s. Since then, the area of robustness has benefited from all mathematical disciplines, from algebra to nonlinear functional analysis, from geometry to variational analysis, which led to many new results. The robustness in turn enriched the mathematics with new nonstandard problems and with their nontraditional solutions.

The last decade has witnessed a big growth of research literature in the area of robustness and nonparametric statistics. High-speed computers enable us to utilize procedures which were earlier considered as purely theoretical. Great progress is observed in multivariate analysis; the extensive datasets demand high-dimensional statistical models, and they further need a rigorous mathematical analysis. Great attention is devoted to the spatial statistical models and to the inference under spatial restrictions. Asymptotics is successively being considered as an aid rather than as an ultimate goal of the research.

Noticing this development, we tried to update some parts of *Robust Statistical Methods with R* to suit the current development. Chapter 5 on multivariate models is extended with multivariate regression models, with invariance and equivariance, and with multivariate two-sample nonparametric tests. Chapter 6, on large samples, is now converted into the large sample and finite sample chapter, illustrating that an estimator with good asymptotic behavior that is asymptotically admissible, can be non-admissible under any finite number of observations. Attention is also devoted to the Newton-Raphson iterative procedure and to the adaptive convex combinations of two estimating procedures. The new Chapter 7, replacing the chapter on goodness-of-fit testing, is devoted to robust procedures in measurement error models.

A great part of the text is devoted to computational aspects. The authors' own procedures can be found in our website: `http://robust.tul.cz/`,

Jana Jurečková, Jan Picek, Martin Schindler
Prague and Liberec, Czech Republic
December, 2018

Preface to the 1st edition

Robust statistical procedures became a part of the general statistical consciousness. Yet, students first learn descriptive statistics and the classical statistical procedures. Only later students and practical statisticians hear that the classical procedures should be used with great caution and that their favorite and simple least squares estimator and other procedures should be replaced with the robust or nonparametric alternatives. To be convinced to use the robust methods, one needs a reasonable motivation; but everybody needs motivation of their own: a mathematically–oriented person demands a theoretical proof, while a practitioner prefers to see the numerical results. Both aspects are important.

The robust statistical procedures became known to the Prague statistical society by the end of the 1960s, thanks to Jaroslav Hájek and his contacts with Peter J. Huber, Peter J. Bickel and other outstanding statisticians. Frank Hampel presented his "Some small sample asymptotics," also touching M-estimates, at Hájek's conference in Prague in 1973; and published his paper in the *Proceedings*. Thus, we had our own experience with the first skepticism toward the robust methods, but by 1980 we started organizing regular workshops for applied statisticians called ROBUST. The 14th ROBUST will be in January 2006. On this occasion, we express our appreciation to Jaromír Antoch, the main organizer.

The course "Robust Statistical Methods" is now a part of the master study of statistics at Charles University in Prague and is followed by all statistical students. The present book draws on experience obtained during these courses. We supplement the theory with examples and computational procedures in the system R.

We chose R as a suitable tool because R seems to be one of the best statistical environments available. It is also free and the R project is an open source project. The code you are using is always available for you to see. Detailed information about the system R and the R project is available from http://www.r-project.org/. The prepared procedures and dataset, not available from the public resource, can be found on website: http://www.fp.vslib.cz/kap/picek/robust/.

We acknowledge the support of the Czech Republic Grant 201/05/2340, the Research Projects MSM 0021620839 of Charles University in Prague, and MSM 467488501 of Technical University in Liberec.

The authors would also like to thank the editors and anonymous referees who contributed considerably to the readability of the text. Our gratitude also belongs to our families for their support and patience.

Jana Jurečková
Jan Picek
Prague and Liberec

Acknowledgments

The present textbook follows up the robust statistical procedures and the relevant mathematical methods, elaborated in an earlier book by the first two authors (2006). During our long cooperation, we learned much from each other and also from a number of excellent researchers and academic groups around the world. We feel a deep gratitude to all these colleagues and also to our students, to whom we tried to forward our knowledge. We are grateful to Martin Schindler, who later joined us with fresh ideas of the young generation. A greater part of our collaborative research was supported by research grants of the Czech Science Foundation GAČR 201/09/0133, 201-12-0083, 15-00243S, 18-01137S and by the Hájek Center for Theoretical and Applied Statistics LC06024. We also acknowledge the support of local facilities in our working places (Charles University in Prague, Technical University of Liberec and the Institute of Information Theory and Automation of the Czech Academy of Sciences (UTIA)).

We thank Rob Calver (Commissioning Editor, Chapman & Hall/CRC Press) for his inspiration and for his continual support to start and to continue in our work, and to all his colleagues who helped us to bring this project to a successful completion.

We thank the "R Core Team" for an excellent software and all the authors of the R packages we use. We thank the reviewers for their careful reading and for constructive comments and suggestions. Last but not least, we thank our spouses Josef, Helena and Lilija for their support and patience during our work on this project.

Introduction

If we analyze the data with the aid of classical statistical procedures, based on parametric models, we usually tacitly assume that the regression is linear, the observations are independent and homoscedastic, and assume the normal distribution of errors. However, when today we can simulate data from any probability distribution and from various models with our high–speed computers and follow the graphics, which was not possible before, we observe that these assumptions are often violated. Then we are mainly interested in the two following questions:

  a) When should we still use the classical statistical procedures, and when are they still optimal?

  b) Are there other statistical procedures that are not so closely connected with special models and conditions?

The classical procedures are typically parametric: the model is fully specified up to the values of several scalar or vector parameters. These parameters typically correspond to the probability distributions of the random errors of the model. If we succeed in estimating these parameters or in testing a hypothesis on their domain, we can use our data and reach a definite conclusion. However, this conclusion is correct only under the validity of our model.

An opposite approach is using the *nonparametric procedures*. These procedures are independent of, or only weakly dependent on, the special shape of the basic probability distribution, and they behave reasonably well (though not just optimally) for a broad class of distribution functions, e.g., that of distribution functions with densities that are eventually symmetric. The discrete probability distributions do not create a serious problem, because their forms usually follow the character of the experiment. A typical representative of nonparametric statistical procedures is the class of the *rank tests of statistical hypotheses*: the "null distributions" of the test criterion (i.e., the distributions under the hypothesis H_0 of interest) coincide under all continuous probability distribution functions of the observations.

Unlike the parametric models with scalar or vector parameters, the nonparametric models consider the whole density function or the regression function as an unknown parameter of infinite dimension. If this functional parameter is only a nuisance, i.e., if our conclusions concern other entities of the model, then we try to avoid its estimation, if possible. The statistical procedures — considering the unknown density, regression function or the influence function as a nuisance parameter, while the inference concerns something else — are known as *semiparametric procedures*; they were developed mainly during the past 30 years. On the other hand, if just the unknown density, regression functions, etc., are our main interests, then we try to find their best possible estimates or the tests of hypotheses on their shapes (goodness-of-fit tests).

Unlike the nonparametric procedures, the *robust statistical procedures* do not try to behave necessarily well for a broad class of models, but they are optimal in some way in a neighborhood of some probability distribution, e.g., normal. Starting with the 1940s, statisticians successively observed that even small deviations from the normal distribution could be harmful, and can strongly impact the quality of the popular least squares estimator, the classical F-test and of other classical methods. Hence, the robust statistical procedures were developed as modifications of the classical procedures, which do not fail under small deviations from the assumed conditions. They are optimal, in a specified sense, in a special neighborhood of a fixed probability distribution, defined with respect to a specified distance of probability measures. As such, the robust procedures are more efficient than the nonparametric ones that pay for their universality by some loss of their efficiency.

When we speak of robust statistical procedures, we usually have estimation procedures in mind. There are also robust tests of statistical hypotheses, namely tests of the Wald type, based on the robust estimators; but whenever possible we recommend using the rank tests instead, mainly for their simplicity and high efficiency.

In the list of various statistical procedures, we cannot omit the *adaptive procedures* that tend to the optimal parametric estimator or test with an increasing number of observations, either in probability or almost surely. Hence, these procedures adapt themselves to the pertaining parametric model with an increasing number of observations, which would be highly desirable. However, this convergence is typically very slow and the optimality is attained under a nonrealistic number of observations. *Partially adaptive procedures* also exist that successively tend to be best from a prescribed finite set of possible decisions; the choice of prescribed finite set naturally determines the success of our inference.

The adaptive, nonparametric, robust and semiparametric methods developed successively, mainly since the 1940s, and they continue to develop as the robust procedures in multivariate statistical models. As such, they are not disjoint from each other, there are no sharp boundaries between these classes, and some concepts and aspects appear in all of them. This book, too, though oriented mainly to robust statistical procedures, often touches other statistical methods. Our ultimate goal is to show and demonstrate which alternative procedures to apply when we are not sure of our model.

Mathematically, we consider the robust procedures as the statistical functionals, defined on the space of distribution functions, and we are interested in their behavior in a neighborhood of a specific distribution or a model.

This neighborhood is defined with respect to a specified distance; hence, we should first consider possible distances on the space of distribution functions and pertaining to basic characteristics of statistical functionals, such as their continuity and derivatives. This is the theoretic background of the robust statistical procedures. However, robust procedures were developed as an alternative to the practical statistical procedures, and they should be applied to practical problems. Keeping this in mind, we also must pay great attention to the computational aspects, and refer to the computational programs that are available or provide our own. As such, we hope that the readers will use, with understanding, robust procedures to solve their problems.

R environment

We chose R as a suitable tool for numerical illustration and computation. R is an open source programming language and environment for statistical analyses and graphics created by Ross Ihaka and Robert Gentleman (1996). It is freely distributed under the terms of the GNU, General Public Licence; since mid-1997 its development and distribution have been carried out by several statisticians known as the "R Core Team." It allows the user

- to store and manipulate data effectively,
- to write their own functions in a user-friendly programming language using loops, conditionals, and recursions;
- to operate on various data structures (like vectors, matrices, etc.); and
- to use the integrated tools for data analysis and graphical display and much more.

R is available in several forms. It compiles and runs on various platforms (UNIX, Windows, MacOS). The files needed to install the standard R GUI, either from the sources (written mainly in C, some routines in Fortran) or from the pre-compiled binaries, are distributed from the Internet site of the Comprehensive R Archive Network (CRAN), see the web page https://www.r-project.org/, where the instructions for the installation ("R Installation and Administration" manual) are also available.

The functionality of R can be extended by installing and loading a package, that is a collection of functions and datasets. There are several "standard" packages that come with the R environment and there are many more available on CRAN.

The R project website also contains a number of tutorials, documents and books that can help one learn R. The manual "An Introduction to R" by the R core development team is an excellent introduction to R for people familiar with statistics. It has many interesting examples of R and a comprehensive treatment of the features of R. Beside that, there is a huge R community providing support through a mailing list or websites with R forums.

To start the R program under Unix, use the command 'R'. Under Windows run, e.g., Rgui.exe, a console-based graphical user interface (GUI). When you use the R program, it issues a prompt when it expects input commands. The default prompt is '>'. In the R GUI, one-line commands can be executed by hitting return. By the assignment operator '<-', the result of an analysis can be saved in a object. If you want to perform a sequence of commands, it is better to create an R script. In that case using an integrated development environment (IDE) for R can be very useful.

RStudio

Our preferred IDE for running R programs is *RStudio*, founded by J.J. Allaire. This software has a free license as well. The first public beta version was released in 2011 and since then it has become the most popular IDE for R. Installation files for major platforms and the source code can be obtained from the web page `https://www.rstudio.com/`. In order to work, RStudio requires a recent installation of R on the system. It has a script editor and visualization and debugging tools. Its four-pane layout makes the work with R easier and more intuitive.

- The *Source pane* displays each file or data object in a separate tab. It allows you to send the current line or selection or the whole content of the active file to the console.

- In the *Console pane* the code is executed by R. It can be entered directly, sent from the source pane or the history sub-pane.

- The *Environment pane* shows a list of objects that are loaded in the workspace including their name, class and dimension. In a separate tab there is also a history of the commands sent to the console.

- The *File and Plot pane* includes a file browser; a Plots tab that enables control over plots created in a session; a Packages tab where packages can be installed, loaded, updated or removed; and a Help tab with links to help and support files.

RStudio has many other nice features like real-time code checking, error detection, code highlighting, auto-completion of function names, small pop-up windows with help or error messages, many keyboard shortcuts, etc.

Chapter 1

Mathematical tools of robustness

1.1 Statistical model

A random experiment leads to some observed values; denote them $X_1, \ldots, X_n$. To make a formal analysis of the experiment and its results, we include everything in the frame of a statistical model. The classical statistical model assumes that the vector $\mathbf{X} = (X_1, \ldots, X_n)$ can attain the values in a *sample space* $\mathcal{X}$ (or $\mathcal{X}_n$) and the subsets of $\mathcal{X}$ are *random events* of our interest. If $\mathcal{X}$ is finite, then there is no problem working with the family of all its subsets. However, some space $\mathcal{X}$ can be too rich, as, e.g., the n-dimensional Euclidean space; then we do not consider all its subsets, but restrict our considerations only to some properly selected subsets/events. In order to describe the experiments and the events mathematically, we consider the family of events that creates a σ-field, i.e., that is closed with respect to the countable unions and the complements of its elements. Let us denote it as $\mathcal{B}$ (or $\mathcal{B}_n$). The probabilistic behavior of random vector $\mathbf{X}$ is described by the *probability distribution* P, which is a set function defined on $\mathcal{B}$. The classical statistical model is a family $\mathcal{P} = \{P_\theta, \theta \in \mathbf{\Theta}\}$ of probability distributions, to which our specific distribution P also belongs. While we can observe $X_1, \ldots, X_n$, the parameter θ is unobservable. It is a real number or a vector and can take on any value in the parametric space $\mathbf{\Theta} \subseteq \mathbb{R}^p$, where p is a positive integer.

The triple $\{\mathcal{X}, \mathcal{B}, P_\theta : \theta \in \mathbf{\Theta}\}$ is the parametric statistical model. The components $X_1, \ldots, X_n$ are often independent copies of a random variable X, but they can also form a segment of a time series. The model is multivariate, when the components X_i, $i = 1, \ldots, n$ are themselves vectors in $\mathbb{R}^k$ with some positive integer k.

The type of experiment often fully determines the character of the parametric model. We easily recognize a special discrete probability distribution, as Bernoulli (alternative distribution), binomial, multinomial, Poisson and hypergeometric distributions.

Binomial distribution: The binomial random variable is a number of successful trials among n independent Bernoulli trials: The i-th Bernoulli trial can result either in success with probability p: then we put $X_i = 1$, or in a failure with probability $1 - p$: then we put $X_i = 0$. In the case of n trials, the binomial random variable X is equal to $X = \sum_{i=1}^{n} X_i$ and can take all integer values

$0, 1, \ldots, n$; specifically,

$$P(X = k) = \binom{n}{k} p^k (1 - p)^{n-k}, \quad k = 0, \ldots, n, \quad 0 \leq p \leq 1$$

The computation can be performed by R function dbinom, for instance:

Example 1.1 $P(X = 3)$: *binomial distribution with $n = 20$, $p = 0.1$*

```
> dbinom(x=3, size=20, prob=0.1)
[1] 0.1901199
```

Multinomial distribution: Let us have n independent trials, each of them leading exactly to one of k different outcomes, to the i-th one with probability p_i, $i = 1, \ldots, k$, $\sum_{i=1}^{k} p_i = 1$. Then the i-th component X_i of the multinomial random vector $\mathbf{X}$ is the number of trials, leading to the outcome i, and

$$P(X_1 = x_1, \ldots, X_n = x_n) = \frac{n!}{x_1! \ldots x_k!} p_1^{x_1} \ldots p_k^{x_k}$$

for any vector $(x_1, \ldots, x_k)$ of integers, $0 \leq x_i \leq n$, $i = 1, \ldots, k$, satisfying $\sum_{i=1}^{k} x_i = n$.

Poisson distribution: The Poisson random variable X is, e.g., the number of clients arriving in the system in a unit interval, the number of electrons emitted from the cathode in a unit interval, etc. Then X can take on all nonnegative integers, and if λ is the intensity of arrivals, emission, etc., then

$$P(X = k) = e^{-\lambda} \frac{\lambda^k}{k!}, \quad k = 0, 1, \ldots$$

Example 1.2 $P(X \leq 5)$: *Poisson distribution with $\lambda = 2$*

```
> ppois(q=5, lambda=2)
[1] 0.9834364
```

Hypergeometric distribution: The hypergeometric random variable X is, e.g., a number of defective items in a sample of size n taken from a finite set of products. If there are M defectives in the set of N products, then

$$P(X = k) = \frac{\binom{M}{k}\binom{N-M}{n-k}}{\binom{N}{n}},$$

for all integers k satisfying $0 \leq k \leq M$ and $0 \leq n - k \leq N - M$; this probability is equal to 0 otherwise. Among the continuous probability distributions, characterized by the densities, we most easily identify the asymmetric distributions concentrated on a halfline. For instance, the waiting time or the duration of a service can be characterized by the gamma distribution.

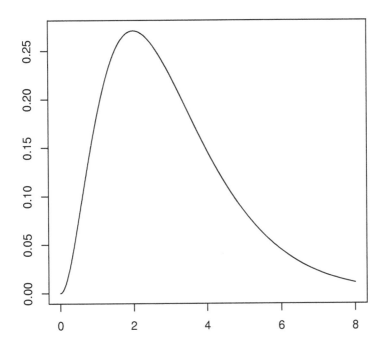

Figure 1.1 *The density function of a gamma distribution with $b = 3$ and $a = 1$.*
Gamma distribution: The gamma random variable X has density function
(see Figure 1.1)

$$f(x) = \begin{cases} \frac{a^b x^{b-1} e^{-ax}}{\Gamma(b)} & \text{if } x \geq 0 \\ 0 & \text{if } x < 0 \end{cases}$$

where a and b are positive constants. The special case $(b=1)$ is the exponential
distribution.

Example 1.3 *The following command plots the density function of a gamma*
distribution with $b = 3$ and $a = 1$ (see Figure 1.1).

```
> plot(seq(0,8,by=0.01),dgamma(seq(0,8,by=0.01),3,1),
+ type="l", ylab="",xlab="")
```

We can generate a sample of size 1000 from the gamma distribution ($b = 3$;
$a = 1$) by

```
> rgamma(1000,3,1)
```
The histogram can be plotted by the R function `hist`.
```
> hist(rgamma(1000,3,1), prob=TRUE,nclass=16)
```
We obtain Figure 1.2, compare with Figure 1.1.

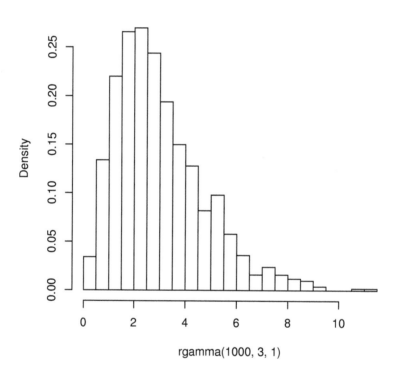

Histogram of rgamma(1000, 3, 1)

Figure 1.2 *The histogram of the simulated sample from the gamma distribution with* $b = 3$ *and* $a = 1$.

In technical practice we can find many other similar examples. However, the symmetric distributions with continuous probability densities are hardly distinguished from each other by simply looking at the data. The problem is also with asymmetric distributions extended on the whole real line. We should either test a hypothesis on their shape or, lacking knowledge of the distribution shape, use a robust or nonparametric method of inference. Most of the statistical procedures elaborated in the past were derived under the normality assumption, i.e., under the condition that the observed data come from a population with the Gaussian/normal distribution. People believed that every symmetric probability distribution described by a density is approximately normal.

Normal distribution: The normal random variable X has density function

$$f(x) = \frac{1}{\sigma\sqrt{2\pi}} e^{-\frac{(x-\mu)^2}{2\sigma^2}}, \quad x \in \mathbb{R}$$

where $\mu \in \mathbb{R}$ and $\sigma^2 > 0$ are constants.

Example 1.4 *95 %-quantile of standard normal distribution*

```
> qnorm(0.95)
[1] 1.644854
```

The procedures based on the normality assumption usually have a simple algebraic structure, thus one is tempted to use them in all situations in which the data can take on symmetrically all real values, forgetting the original normality assumption. For instance, the most popular least squares estimator (LSE) of regression or other parameters, though seemingly universal, is closely connected with the normal distribution of the measurement errors. That itself would not matter, but the LSE fails when even a small fraction of data comes from another population whose distribution has heavier tails than the normal one, or when the dataset is contaminated by some outliers.

At present, these facts can be easily demonstrated numerically with simulated data, while this was not possible before the era of high-speed computers. But these facts are not only verified with computers; the close connection of the least squares and the normal distribution is also supported by strong theoretical arguments, based on the characterizations of the normal distribution by means of properties of estimators and other procedures. For instance, Kagan, Linnik, and Rao (1973) proved that the least squares estimator (LSE) of the regression parameter in the linear regression model with a continuous distribution of the errors is admissible with respect to the quadratic risk (i.e., there is no other estimator with uniformly better quadratic risk), if and only if the distribution of the measurement errors is normal.

The Student t-test, the Snedecor F-test and the test of the linear hypothesis were derived under the normality assumption. While the t-test is relatively robust to deviations from the normality, the F-test is very sensitive in this sense and should be replaced with a rank test, unless the normal distribution is taken for granted.

If we are not sure by the parametric form of the model, we can use either of the following possible alternative procedures:

a) *Nonparametric approach:* We give up a parametrization of P_θ by a real or vector parameter θ, and replace the family $\{P_\theta : \theta \in \Theta\}$ with a broader family of probability distributions.

b) *Robust approach:* We introduce an appropriate measure of distance of statistical procedures made on the sample space $\mathcal{X}$, and study the stability of the classical procedures, optimal for the model P_θ, under small deviations from this model. At the same time, we try to modify slightly the classical procedures (i.e., to find robust procedures) to reduce their sensitivity.

1.2   Illustration on statistical estimation

Let $X_1, \ldots, X_n$ be independent observations, identically distributed with some probability distribution P_θ, where θ is an unobservable parameter, $\theta \in \Theta \subseteq \mathbb{R}^p$. Let $F(x, \theta)$ be the distribution function of P_θ. Our problem is to estimate the unobservable parameter θ.

We have several possibilities, for instance

(1) maximal likelihood method,

(2) moment method,

(3) method of χ^2-minimum, or another method minimizing another distance of the empirical and the true distributions, or

(4) method based on the sufficient statistics (Rao-Blackwell Theorem) and on the complete sufficient statistics (Lehmann-Scheffé Theorem).

In the context of sufficient statistics, remember the very useful fact in nonparametric models that the ordered sample (the vector of order statistics) $X_{n:1} \leq X_{n:2} \leq \ldots \leq X_{n:n}$ is a complete sufficient statistic for the family of probability distributions with densities $\prod_{i=1}^n f(x_i)$, where f is an arbitrary continuous density. This corresponds to the model in which the observations create an independent random sample from an arbitrary continuous distribution. If θ is a one-dimensional parameter, thus a real number, we are intuitively led to the class of L-estimators of the type

$$T_n = \sum_{i=1}^n c_{ni} h(X_{n:i})$$

based on order statistics, with suitable coefficients c_{ni}, $i = 1, \ldots, n$, and a suitable function $h(\cdot)$.

(5) Minimization of some (criterion) function of observations and of θ: e.g., the minimization

$$\sum_{i=1}^n \rho(X_i, \theta) := \min, \quad \theta \in \Theta$$

with a suitable non-constant function $\rho(\cdot, \cdot)$. As an example we can consider $\rho(x, \theta) = -\log f(x, \theta)$ leading to the maximal likelihood estimator $\hat{\theta}_n$. The estimators of this type are called M-estimators, or estimators of the maximum likelihood type.

(6) An inversion of the rank tests of the shift in location, of the significance of regression, etc., leads to the class of R-estimators, based on the ranks of the observations or of their residuals.

These are the M-, L- and R-estimators, and some other robust methods that create the main subject of this book.

1.3 Statistical functional

Consider a random variable X with probability distribution P_θ with distribution function F, where $P_\theta \in \mathcal{P} = \{P_\theta : \theta \in \Theta \subseteq \mathbb{R}^p\}$. Then in many cases θ can be looked at as a functional $\theta = T(P)$ defined on $\mathcal{P}$; we can also write $\theta = T(F)$. Intuitively, a natural estimator of θ, based on observations $X_1, \ldots, X_n$, is $T(P_n)$, where P_n is *the empirical probability distribution* of vector $(X_1, \ldots, X_n)$, i.e.,

$$P_n(A) = \frac{1}{n} \sum_{i=1}^n I[X_i \in A], \quad A \in \mathcal{B} \tag{1.1}$$

Otherwise, P_n is the uniform distribution on the set $\{X_1, \ldots, X_n\}$, because $P_n(\{X_i\}) = \frac{1}{n}$, $i = 1, \ldots, n$. The distribution function, pertaining to P_n, is the *empirical distribution function*

$$F_n(x) = P_n((-\infty, x]) = \frac{1}{n} \sum_{i=1}^n I[X_i \le x], \; x \in \mathbb{R} \tag{1.2}$$

Here are some examples of functionals:

(1) Expected value:

$$T(P) \quad = \int_{\mathbb{R}} x \, dP \quad = \mathbb{E}X$$

$$T(P_n) \quad = \int_{\mathbb{R}} x \, dP_n \quad = \bar{X}_n = \frac{1}{n} \sum_{i=1}^n X_i$$

(2) Variance:

$$T(P) \quad = \quad \mathrm{var}\, X = \int_{\mathbb{R}} x^2 \, dP - (\mathbb{E}X)^2$$

$$T(P_n) \quad = \quad \frac{1}{n} \sum_{i=1}^n X_i^2 - \bar{X}_n^2$$

(3) If $T(P) = \int_{\mathbb{R}} h(x) \, dP$, where h is an arbitrary P-integrable function, then an empirical counterpart of $T(P)$ is

$$T(P_n) = \frac{1}{n} \sum_{i=1}^n h(X_i)$$

(4) Conversely, we can find a statistical functional corresponding to a given statistical estimator: for instance, the *geometric mean* of observations $X_1, \ldots, X_n$ is defined as

$$T(P_n) = G_n = \left(\prod_{i=1}^n X_i \right)^{1/n}$$

$$\log G_n = \frac{1}{n}\sum_{i=1}^{n}\log X_i = \int_{\mathbb{R}}\log x \, dP_n$$

hence the corresponding statistical functional has the form

$$T(P) = \exp\left\{\int_{\mathbb{R}}\log x \, dP\right\}$$

Similarly, the *harmonic mean* $T(P_n) = H_n$ of observations $X_1,\ldots,X_n$ is defined as

$$\frac{1}{H_n} = \frac{1}{n}\sum_{i=1}^{n}\frac{1}{X_i}$$

and the corresponding statistical functional has the form

$$T(P) = H = \left(\int_{\mathbb{R}}\frac{1}{x}dP\right)^{-1}$$

Statistical functionals were first considered by von Mises (1947).

The estimator $T(P_n)$ should tend to $T(P)$, as $n \to \infty$, with respect to some type of convergence defined on the space of probability measures. Mostly, it is a convergence in probability and in distribution, or almost sure convergence; but an important characteristic also is the large sample bias of estimator $T(P_n)$, i.e., $\lim_{n\to\infty}|\mathbb{E}[T(P_n) - T(P)]|$, which corresponds to the convergence in the mean. Because we need to study the behavior of $T(P_n)$ also in a neighborhood of P, we consider an expansion of the functional $(T(P_n) - T(P))$ of the Taylor type. To do it, we need some concepts of the functional analysis, as various distances P_n and P, and their relations, and the continuity and differentiability of functional T with respect to the considered distance.

1.4   Fisher consistency

A reasonable statistical estimator should have the natural property of Fisher consistency, introduced by R. A. Fisher (1922). We say that estimator $\hat{\theta}_n$, based on observations $X_1 \ldots, X_n$ with probability distribution P, is a Fisher consistent estimator of parameter θ, if, written as a functional $\hat{\theta}_n = T(P_n)$ of the empirical probability distribution of vector $(X_1,\ldots,X_n)$, $n = 1,\ldots$, it satisfies $T(P) = \theta$. The following example shows that this condition is not always automatically satisfied.

- Let $\theta = \mathrm{var}\ X = T(P) = \int_{\mathbb{R}} x^2 dP - \left(\int_{\mathbb{R}} x dP\right)^2$ be the variance of P. Then the sample variance $\hat{\theta}_n = T(P_n) = \frac{1}{n}\sum_{i=1}^{n}(X_i - \bar{X}_n)^2$ is Fisher consistent; but it is biased, because $\mathbb{E}\hat{\theta}_n = \left(1 - \frac{1}{n}\right)\theta$. On the other hand, the unbiased estimator of the variance $S_n^2 = \frac{1}{n-1}\sum_{i=1}^{n}(X_i - \bar{X}_n)^2$ is not a Fisher consistent estimator of θ, because

$$S_n^2 = \frac{n}{n-1}T(P_n) \quad \text{and} \quad \frac{n}{n-1}T(P) \neq T(P)$$

From the robustness point of view, the natural property of Fisher consistency of an estimator is more important than its unbiasedness; hence it should be first checked on a statistical functional.

1.5   Some distances of probability measures

Let $\mathcal{X}$ be a metric space with metric d, separable and complete, and denote $\mathcal{B}$ as the σ-field of its Borel subsets. Furthermore, let $\mathcal{P}$ be the system of all probability measures on the space $(\mathcal{X}, \mathcal{B})$. Then $\mathcal{P}$ is a convex set, and on $\mathcal{P}$ we can introduce various distances of its two elements $P, Q \in \mathcal{P}$.

Let us briefly describe some such distances, mostly used in mathematical statistics. For those who want to learn more about these and other distances and other related topics, we refer to the literature of the functional analysis and the probability theory, e.g., Billingsley (1998) or Fabian et al. (2001).

(1) *The Prochorov distance:*

$$d_P(P, Q) \quad = \quad \inf\{\varepsilon > 0: \ P(A) \le Q(A^\varepsilon) + \varepsilon$$
$$\forall A \in \mathcal{B}, A \ne \emptyset\}$$

where $A^\varepsilon = \{x \in \mathcal{X}: \ \inf_{y \in A} d(x, y) \le \varepsilon\}$ is a closed ε-neighborhood of a non-empty set A.

(2) *The Lévy distance:* $\mathcal{X} = \mathbb{R}$ is the real line; let F, G be the distribution functions of probability measures P, Q, then

$$d_L(F, G) \quad = \quad \inf\{\varepsilon > 0: \ F(x - \varepsilon) - \varepsilon$$
$$\le G(x) \le F(x + \varepsilon) + \varepsilon \forall x \in \mathbb{R}\}$$

(3) *The total variation:*

$$d_V(P, Q) = \sup_{A \in \mathcal{B}} |P(A) - Q(A)|$$

We easily verify that $d_V(P, Q) = \int_{\mathcal{X}} |dP - dQ|$

(4) *The Kolmogorov distance:* $\mathcal{X} = \mathbb{R}$ is the real line and F, G are the distribution functions of probability measures P, Q, then

$$d_K(F, G) = \sup_{x \in \mathbb{R}} |F(x) - G(x)|$$

(5) *The Hellinger distance:*

$$d_H(P, Q) = \left\{ \int_{\mathcal{X}} \left(\sqrt{dP} - \sqrt{dQ} \right)^2 \right\}^{1/2}$$

If $f = \frac{dP}{d\mu}$ and $g = \frac{dQ}{d\mu}$ are densities of P, Q with respect to some measure μ, then the Hellinger distance can be rewritten in the form

$$(d_H(P,Q))^2 = \int_{\mathcal{X}} \left(\sqrt{f} - \sqrt{g} \right)^2 d\mu = 2 \left(1 - \int_{\mathcal{X}} \sqrt{fg} d\mu \right)$$

(6) *The Lipschitz distance:* Assume that $d(x, y) \le 1 \; \forall x, y \in \mathcal{X}$ (we take the metric $d' = \frac{d}{1+d}$ otherwise), then

$$d_{Li}(P,Q) = \sup_{\psi \in \mathcal{L}} \left| \int_{\mathcal{X}} \psi dP - \int_{\mathcal{X}} \psi dQ \right|$$

where $\mathcal{L} = \{\Psi : \mathcal{X} \mapsto \mathbb{R} : |\psi(x) - \psi(y)| \le d(x, y)\}$ is the set of the Lipschitz functions.

(7) *Kullback-Leibler divergence:* Let p, q be the densities of probability distributions P, Q with respect to measure μ (Lebesgue measure on the real line or the counting measure), then

$$d_{KL}(Q,P) = \int q(x) \ln \frac{q(x)}{p(x)} d\mu(x)$$

The Kullback-Leibler divergence is not a metric, because it is not symmetric in P, Q and does not satisfy the triangle inequality.

More on distances of probability measures can be found in Gibbs and Su (2002), Liese and Vajda (1987), Rachev (1991), Reiss (1989) and Zolotarev (1983), among others.

Example 1.5 *We can use R to find some of the distances of two discrete distributions. Let P be the binomial distribution with parameters $n = 100$, $p = 0.01$. Let Q be Poisson distribution with parameter $\lambda = np = 1$. Then*

$$2d_V(P,Q) = \sum_{k=0}^{100} \left| \binom{n}{k} 0.01^k 0.99^{100-k} - \frac{e^{-1}}{k!} \right| + \sum_{k=101}^{\infty} \frac{e^{-1}}{k!}$$

```
> sum(abs(choose(100,0:100)*0.01^(0:100)*(0.99)^(100:0)
+ -exp(-1)/factorial(0:100)))+1-sum(exp(-1)/factorial(0:100))
[1] 0.005550589

> ## or also
> sum(abs(dbinom(0:100,100,0.01)-dpois(0:100,1)))+1-ppois(100,1)
[1] 0.005550589
```

Thus $d_V(Bi(100, 0.01), Po(1)) \approx 0.0028$. Similarly, $d_K(Bi(100, 0.01), Po(1)) \approx 0.0018$, $d_H(Bi(100, 0.01), Po(1)) \approx 0.0036$, $d_{KL}(Bi(100, 0.01), Po(1)) \approx 0.000025$ because

```
> ### Kolmogorov distance
> max(abs(pbinom(0:100,100,0.01)-ppois(0:100,1)))
[1] 0.0018471

> ### Hellinger distance
> sqrt(sum((sqrt(dbinom(0:100,100,0.01))
+  -sqrt(dpois(0:100,1)))^2))
[1] 0.003562329

>### Kullback-Leibler divergence (Q,P)
> sum(dpois(0:100,1)*log(dpois(0:100,1)/
+ dbinom(0:100,100,0.01)))
[1] 2.551112e-05

>### Kullback-Leibler divergence (P,Q)
> sum(dbinom(0:100,100,0.01)*log(dbinom(0:100,100,0.01)/
+   dpois(0:100,1)))
[1] 2.525253e-05
```

1.6   Relations between distances

The family $\mathcal{P}$ of all probability measures on $(\mathcal{X}, \mathcal{B})$ is a metric space with respect to each of the distances described above. On this metric space we can study the continuity and other properties of the statistical functional $T(P)$. Because we are interested in the behavior of the functional, not only at distribution P, but also in its neighborhood; we come to the question, which distance is more sensitive to small deviations of P?

The following inequalities between the distances show not only which distance eventually dominates above others, but also illustrate their relations. Their verification we leave as an exercise:

$$d_H^2(P,Q) \quad \le 2d_V(P,Q) \quad \le 2d_H(P,Q)$$

$$d_P^2(P,Q) \quad \le d_{Li}(P,Q) \quad \le 2d_P(P,Q) \quad \forall\ P,Q \in \mathcal{P} \qquad (1.3)$$

$$\frac{1}{2}d_V^2(P,Q) \quad \le d_{KL}(P,Q)$$

if $\mathcal{X} = \mathbb{R}$, then it further holds:

$$d_L(P,Q) \quad \le d_P(P,Q) \quad \le d_V(P,Q)$$

$$d_L(P,Q) \quad \le d_K(P,Q) \quad \le d_V(P,Q) \quad \forall\ P,Q \in \mathcal{P} \qquad (1.4)$$

Example 1.6 *Let P be the exponential distribution with density*

$$f(x) = \begin{cases} e^{-x} & \dots & x \ge 0 \\ 0 & \dots & x < 0 \end{cases}$$

and let Q be the uniform distribution $R(0, 1)$ with density

$$g(x) = \begin{cases} 1 & \dots & 0 \le x \le 1 \\ 0 & \dots & otherwise \end{cases}$$

Then

$$2d_V(P, Q) = \int_0^1 \left(1 - e^{-x}\right) dx + \int_1^\infty e^{-x} dx = 1 + \frac{1}{e} - 1 + \frac{1}{e} = \frac{2}{e}$$

hence $d_V(\exp, R(0, 1)) \approx 0.3679$. For comparison we can use the R function `integrate` *to compute the integral. Compare the following result with the calculation above.*

```
> integrate(function(x) {1-exp(-x)},0,1)
0.3678794 with absolute error < 4.1e-15
> integrate(function(x) {exp(-x)},1, Inf)
0.3678794 with absolute error < 2.1e-05
```

Furthermore,

$$\begin{aligned} d_K(P, Q) &= \sup_{x \ge 0} \left|1 - e^{-x} - xI[0 \le x \le 1] - I[x > 1]\right| \\ &= e^{-1} \approx 0.1839 \end{aligned}$$

and

$$d_H^2(\exp, R(0, 1)) = 2 \left(1 - \int_0^1 \sqrt{e^{-x}} dx\right) = 2 \left(\frac{2}{\sqrt{e}} - 1\right)$$

thus $d_H(\exp, R(0, 1)) \approx 0.6528$. By numerical integration we get

```
> sqrt(2*(1-integrate(function(x) {sqrt(exp(-x))},0,1)$value))
[1] 0.6527807
```

Finally

$$d_{KL}(R(0, 1), \exp) = \int_0^1 \ln \frac{1}{e^{-x}} dx = \frac{1}{2}$$

1.7   Differentiable statistical functionals

Let $\mathcal{P}$ again be the family of all probability measures on the space $(\mathcal{X}, \mathcal{B}, \mu)$, and assume that $\mathcal{X}$ is a complete separable metric space with metric d and that $\mathcal{B}$ is the system of the Borel subsets of $\mathcal{X}$. Choose some distance δ on $\mathcal{P}$ and consider the statistical functional $T(\cdot)$ defined on $\mathcal{P}$. If we want to analyze an expansion of $T(\cdot)$ around P, analogous to the Taylor expansion, we must introduce the concept of a derivative of statistical functional. There are more possible definitions of the derivative, and we shall consider three of them: the Gâteau derivative, the Fréchet and the Hadamard derivative, and compare their properties from the statistical point of view.

Definition 1.1 *Let $P, Q \in \mathcal{P}$ and let $t \in [0, 1]$. Then the probability distribution*

$$P_t(Q) = (1 - t)P + tQ \tag{1.5}$$

is called the **contamination** *of P by Q in ratio t.*

Remark 1.1 *$P_t(Q)$ is a probability distribution, because $\mathcal{P}$ is convex. $P_0(Q) = P$ means an absence of the contamination, while $P_1(Q) = Q$ means the full contamination.*

1.8   Gâteau derivative

Fix two distributions $P, Q \in \mathcal{P}$ and denote $\varphi(t) = T((1-t)P+tQ)$, $0 \leq t \leq 1$. Suppose that the function $\varphi(t)$ has the final n-th derivative $\varphi^{(n)}$, and that the derivatives $\varphi^{(k)}$ are continuous in interval $(0, 1)$ and that the right-hand derivatives $\varphi_+^{(k)}$ are right-continuous at $t = 0$, $k = 1, \ldots, n - 1$. Then we can consider the Taylor expansion around $u \in (0, 1)$

$$\varphi(t) = \varphi(u) + \sum_{k=1}^{n-1} \frac{\varphi^{(k)}(u)}{k!}(t - u)^k + \frac{\varphi^{(n)}(v)}{n!}(t - u)^n, \; v \in [u, t] \tag{1.6}$$

We are mostly interested in the expansion on the right of $u = 0$, that corresponds to a small contamination of P. For that we replace derivatives $\varphi^{(k)}(0)$ with the right-hand derivatives $\varphi_+^{(k)}(0)$. The derivative $\varphi_+'(0)$ is called the *Gâteau derivative* of functional T in P in direction Q.

Definition 1.2 *We say that functional T is differentiable in the Gâteau sense in P in direction Q, if there exists the limit*

$$T_Q'(P) = \lim_{t \to 0_+} \frac{T(P + t(Q - P)) - T(P)}{t} \tag{1.7}$$

$T_Q'(P)$ is called the Gâteau derivative T in P in direction Q.

Remark 1.2

a)   *The Gâteau derivative $T_Q'(P)$ of functional T is equal to the ordinary right derivative of function φ at the point 0, i.e.,*

$$T_Q'(P) = \varphi'(0_+)$$

b)   *Similarly defined is the Gâteau derivative of order k:*

$$T_Q^{(k)}(P) = \left[\frac{d^k}{dt^k} T(P + t(Q - P)) \right]_{t=0_+} = \varphi^{(k)}(0_+)$$

c)   In the special case when Q is the Dirac probability measure $Q = \delta_x$ assigning probability 1 to the one-point set $\{x\}$ x, we shall use a simpler notation $T'_{\delta_x}(P) = T'_x(P)$

In the special case $t = 1$, $u = 0$ the Taylor expansion (1.6) reduces to the form

$$T(Q) - T(P) = \sum_{k=1}^{n-1} \frac{T_Q^{(k)}(P)}{k!} + \frac{1}{n!}\left[\frac{d^n}{dt^n}T(P + t(Q - p))\right]_{t=t^*} \tag{1.8}$$

where $0 \leq t^* \leq 1$.

As an illustration we consider the following functionals.

(a) Expected value:

$$T(P) = \int_{\mathcal{X}} x dP = \mathbb{E}_P X$$

$$\varphi(t) = \int_{\mathcal{X}} x d((1 - t)P + tQ) = (1 - t)\mathbb{E}_P X + t\mathbb{E}_Q X$$

$$\implies \varphi'(t) = \mathbb{E}_Q X - \mathbb{E}_P X$$

$$T'_Q(P) = \varphi'(0_+) = \mathbb{E}_Q X - \mathbb{E}_P X.$$

Finally we obtain for $Q = \delta_x$

$$T'_x = x - \mathbb{E}_P X$$

(b) Variance:

$$T(P) = \mathrm{var}_P X = \mathbb{E}_P(X^2) - (\mathbb{E}_P X)^2$$

$$T((1 - t)P + tQ) = \int_{\mathcal{X}} x^2 d((1 - t)P + tQ)$$

$$- \left[\int_{\mathcal{X}} x d((1 - t)P + tQ)\right]^2$$

$$\implies \varphi(t) = (1 - t)\mathbb{E}_P X^2 + t\mathbb{E}_Q X^2 - (1 - t)^2(\mathbb{E}_P X)^2$$

$$-t^2 (\mathbb{E}_Q X)^2 - 2t(1 - t)\mathbb{E}_P X \cdot \mathbb{E}_Q X$$

$$\varphi'(t) = -\mathbb{E}_P X^2 + \mathbb{E}_Q X^2$$

$$+2(1 - t)(\mathbb{E}_P X)^2 - 2t (\mathbb{E}_Q X)^2$$

$$-2(1 - 2t)\mathbb{E}_P X \cdot \mathbb{E}_Q X$$

This further implies

$$\lim_{t \to 0_+} \varphi'(t) = T'_Q(P)$$

$$= \mathbb{E}_Q X^2 - \mathbb{E}_P X^2 - 2\mathbb{E}_P X \cdot \mathbb{E}_Q X + 2 (\mathbb{E}_P X)^2$$

and finally we obtain for $Q = \delta_x$

$$
\begin{aligned}
T'_x(P) &= x^2 - \mathbb{E}_P X^2 - 2x\mathbb{E}_P X + 2\left(\mathbb{E}_P X\right)^2 \\
&= (x - \mathbb{E}_P X)^2 - \operatorname{var}_P X
\end{aligned}
\tag{1.9}
$$

1.9   Fréchet derivative

Definition 1.3 *We say that functional T is differentiable in P in the Fréchet sense, if there exists a linear functional $L_P(Q - P)$ such that*

$$
\lim_{t \to 0} \frac{T(P + t(Q - P)) - T(P)}{t} = L_P(Q - P)
\tag{1.10}
$$

uniformly in $Q \in \mathcal{P}$, $\delta(P, Q) \leq C$ for any fixed $C \in (0, \infty)$.
   The linear functional $L_P(Q - P)$ is called the Fréchet derivative of functional T in P in direction Q.

Remark 1.3

a)   Because L_P is a linear functional, there exists a function $g : \mathcal{X} \mapsto \mathbb{R}$ such that

$$
L_P(Q - P) = \int_{\mathcal{X}} g \, d(Q - P)
\tag{1.11}
$$

b)   If T is differentiable in the Fréchet sense, then it is differentiable in the Gâteau sense, too, i.e., there exists $T'_Q(P) \; \forall Q \in \mathcal{P}$, and it holds

$$
T'_Q(P) = L_P(Q - P) \quad \forall Q \in \mathcal{P}
\tag{1.12}
$$

Especially,

$$
T'_x(P) = L_P(\delta_x - P) = g(x) - \int_{\mathcal{X}} g \, dP
\tag{1.13}
$$

and this further implies

$$
\mathbb{E}_P(T'_x(P)) = \int_{\mathcal{X}} T'_x(P) \, dP = 0.
\tag{1.14}
$$

c)   Let P_n be the empirical probability distribution of vector $(X_1 \ldots, X_n)$. Then $P_n - P = \frac{1}{n} \sum_{i=1}^{n} (\delta_{X_i} - P)$. Hence, because L_P is a linear functional,

$$
L_P(P_n - P) = \frac{1}{n} \sum_{i=1}^{n} L_P(\delta_{X_i} - P) = \frac{1}{n} \sum_{i=1}^{n} T'_{X_i}(P) = T'_{P_n}(P)
\tag{1.15}
$$

Proof of (1.12):

Actually, because $L_P(\cdot)$ is a linear functional, we get by (1.10)

$$
\begin{aligned}
T'_Q(P) &= \lim_{t \to 0_+} \frac{T(P + t(Q - P)) - T(P)}{t} \\[2mm]
&= \lim_{t \to 0_+} \frac{T(P + t(Q - P)) - T(P)}{t} - L_P(Q - P) \\[2mm]
&\quad + L_P(Q - P) = 0 + L_P(Q - P) = L_P(Q - P) \qquad \square
\end{aligned}
$$

1.10 Hadamard (compact) derivative

If there exists a linear functional $L(Q - P)$ such that the convergence (1.10) is uniform not necessarily for bounded subsets of the metric space $(\mathcal{P}, \delta)$ containing P, i.e., for all Q satisfying $\delta(P, Q) \le C$, $0 < C < \infty$, but only for Q from any fixed compact set $\boldsymbol{K} \subset \mathcal{P}$ containing P; then we say that functional T is differentiable in the *Hadamard sense*, and we call the functional $L(Q - P)$ the *Hadamard (compact) derivative* of T.

The Fréchet differentiable functional is obviously also Hadamard differentiable, and it is, in turn, also Gâteau differentiable, similar to Remark 1.3. We refer to Fernholz (1983) and to Fabian et al. (2001) for more properties of differentiable functionals.

The Fréchet differentiability imposes rather restrictive conditions on the functional that are not satisfied namely by the robust functionals. On the other hand, when we have a Fréchet differentiable functional, we can easily derive the large sample (normal) distribution of its empirical counterpart, when the number n of observations infinitely increases. If the functional is not sufficiently smooth, we can sometimes derive the large sample normal distribution of its empirical counterpart with the aid of the Hadamard derivative. If we only want to prove that $T(P_n)$ is a consistent estimator of $T(P)$, then it suffices to consider the continuity of $T(P)$.

The Gâteau derivative of $T'_x(P)$, called the *influence function* of functional T, is one of the most important characteristics of its robustness and will be studied in Chapter 2 in detail.

1.11 Large sample distribution of empirical functional

Consider again the metric space $(\mathcal{P}, \delta)$ of all probability distributions on $(\mathcal{X}, \mathcal{B})$, with metric δ satisfying

$$
\sqrt{n}\,\delta(P_n, P) = O_p(1) \quad \text{as } n \to \infty, \tag{1.16}
$$

where P_n is the empirical probability distribution of the random sample $(X_1, \ldots, X_n)$, $n = 1, 2, \ldots$. The convergence (1.16) holds, e.g., for the Kolmogorov distance of the empirical distribution function from the true one, which is the most important for statistical applications; but it holds also for other distances.

As an illustration of the use of the functional derivatives, let us show that the Fréchet differentiability, together with the classical central limit theorem, always give the large sample (asymptotic) distribution of the empirical functional $T(P_n)$.

Theorem 1.1 *Let T be a statistical functional, Fréchet differentiable in P, and assume that the empirical probability distribution P_n of the random sample $(X_1, \ldots, X_n)$ satisfies the condition (1.16) as $n \to \infty$. If the variance of the Gâteau derivative $T'_{X_1}(P)$ is positive, $\operatorname{var}_P T'_{X_1}(P) > 0$, then the sequence $\sqrt{n}(T(P_n) - T(P))$ is asymptotically normally distributed as $n \to \infty$, namely*

$$\mathcal{L}\Big(T(P_n) - T(P)\Big) \longrightarrow \mathcal{N}\Big(0, \operatorname{var}_P T'_{X_1}(P)\Big) \qquad (1.17)$$

Proof. By (1.15), $T'_{P_n}(P) = \frac{1}{n}\sum_{i=1}^{n} T'_{X_i}(P)$ and further by (1.8) and condition (1.16) we obtain

$$\sqrt{n}(T(P_n) - T(P)) = \frac{1}{\sqrt{n}} \sum_{i=1}^{n} T'_{X_i}(P) + R_n$$

$$= \frac{1}{\sqrt{n}} \sum_{i=1}^{n} L_P(P_n - P) + \sqrt{n}\, o(\delta(P_n, P)) \qquad (1.18)$$

$$= \frac{1}{\sqrt{n}} \sum_{i=1}^{n} T'_{X_i}(P) + o_p(1)$$

If the joint variance $\operatorname{var}_P T'_{X_i}(P) = \operatorname{var}_P T'_{X_1}(P)$, $i = 1, \ldots, n$, is finite, then (1.17) follows from (1.18) and from the classical central limit theorem.   □

• *Sample variance:* Let $T(P) = \operatorname{var}_P X = \sigma^2$, then

$$T(P_n) = S_n^2 = \frac{1}{n} \sum_{i=1}^{n} (X_i - \bar{X}_n)^2$$

and, by (1.9)

$$T'_x(P) = (x - \mathbb{E}_P X)^2 - \operatorname{var}_P X$$

hence

$$\operatorname{var}_P T'_X(P) = \mathbb{E}_P (X - \mathbb{E}_P X)^4 - \mathbb{E}_P^2 (X - \mathbb{E}_P X)^2 = \mu_4 - \mu_2^2$$

and by Theorem 1.1 we get the large sample distribution of the sample variance

$$\mathcal{L}\Big(\sqrt{n}(S_n^2 - \sigma^2)\Big) \longrightarrow \mathcal{N}\Big(0, \mu_4 - \mu_2^2\Big)$$

1.12   Problems and complements

1.1 Let Q be the binomial distribution with parameters n, p and let P be the Poisson distribution with parameter $\lambda = np$, then

$$\frac{1}{2}(d_V(Q, P))^2 \le d_{KL}(Q, P)$$

$$\frac{p^2}{4} \le d_{KL}(Q, P) \le \left(\frac{1}{4} + \frac{np^3}{3} + \frac{p}{2} + \frac{1}{4n}\right) p^2$$

$$\frac{1}{16} \min(p, np^2) \le d_V(Q, P) \le 2p \left(1 - e^{-np}\right)$$

$$d_{KL}(Q, P) \le \frac{p^2}{2(1 - p)}$$

$$\lim_{n \to \infty} n^2 d_{KL}(Q, P) = \frac{\lambda^2}{4}$$

See Barbour and Hall (1984), Csiszár (1967), Harremoës and Ruzankin (2004), Kontoyannis et al. (2005), Pinsker (1960) for demonstrations.

1.2 *Wasserstein-Kantorovich distance* of distribution functions F, G of random variables X, Y:

- L_1-distance on $\mathcal{F}_1 = \{F : \int_{-\infty}^{\infty} |x| dF(x) < \infty\}$:

$$d_W^{(1)}(F, G) = \int_0^1 |F^{-1}(t) - G^{-1}(t)| dt$$

Show that $d_W^{(1)}(F, G) = \int_{-\infty}^{\infty} |F(x) - G(x)| dx$ (Dobrushin (1970)).

Show that $d_W^{(1)}(F, G) = \inf\{\mathbb{E}|X - Y|\}$ where the infimum is over all jointly distributed X and Y with respective marginals F and G.

- L_2-distance on $\mathcal{F}_2 = \{F : \int_{-\infty}^{\infty} x^2 dF(x) < \infty\}$:

$$d_W^{(2)}(F, G) = \int_0^1 [F^{-1}(t) - G^{-1}(t)]^2 dt$$

Show that $d_W^{(2)}(F, G) = \inf\{\mathbb{E}(X - Y)^2\}$ where the infimum is over all jointly distributed X and Y with respective marginals F and G (Mallows (1972)).

- Weighted L_1-distance:

$$d_W^{(1)}(F, G) = \int_0^1 |F^{-1}(t) - G^{-1}(t)| w(t) dt, \quad \int_0^1 w(t) dt = 1$$

1.3 Show that $d_P(F, G) \leq d_W^{(1)}(F, G)$ (Dobrushin (1970)).

1.4 Let $(X_1, \ldots, X_n)$ and $(Y_1, \ldots, Y_n)$ be two independent random samples from distribution functions F, G such that $\int_{-\infty}^{\infty} x dF(x) = \int_{-\infty}^{\infty} y dG(y) = 0$. Let F_n, G_n be the distribution functions of $n^{-1/2} \sum_{i=1}^{n} X_i$ and $n^{-1/2} \sum_{i=1}^{n} Y_i$, respectively, then

$$d_W^{(2)}(F_n, G_n) \leq d_W^{(2)}(F, G)$$

(Mallows (1972)).

1.5 χ^2-*distance*: Let p, q be the densities of probability distributions P, Q with respect to measure μ (μ can be a countable measure). Then $d_{\chi^2}(P, Q)$ is defined as

$$d_{\chi^2}(P, Q) = \int_{x \in \mathcal{X}: p(x), q(x) > 0} \frac{(p(x) - q(x))^2}{q(x)} d\mu(x)$$

Then $0 \leq d_{\chi^2}(P, Q) \leq \infty$ and d_{χ^2} is independent of the choice of the dominating measure. It is not a metric, because it is not symmetric in P, Q. Distance d_{χ^2} dates back to Pearson in the 1930s and has many applications in statistical inference. The following relations hold between d_{χ^2} and other distances:

(i) $d_H(P, Q) \leq \sqrt{2}(d_{\chi^2}(P, Q))^{1/4}$

(ii) If the sample space $\mathcal{X}$ is countable, then $d_V(P, Q) \leq \frac{1}{2}\sqrt{d_{\chi^2}(P, Q)}$

(iii) $d_{KL}(P, Q) \leq d_{\chi^2}(P, Q)$

1.6 Let P be the exponential distribution and let Q be the uniform distribution (see Example 1.6) Then

$$d_{\chi^2}(Q, P) = \int_0^1 \frac{(1 - e^{-x})^2}{e^{-x}} dx = e + e^{-1} - 2$$

hence $d_{\chi^2}(R(0, 1), \exp) \approx 0.350402$. Furthermore,

$$d_{\chi^2}(P, Q) = \int_0^1 \left(e^{-x} - 1\right)^2 dx = -\frac{1}{2}e^{-2} + 2e^{-1} - \frac{1}{2}$$

hence $d_{\chi^2}(\exp, R(0, 1)) \approx 0.168091$

1.7 *Bhattacharyya distance*: Let p, q be the densities of probability distributions P, Q with respect to measure. Then $d_B(P, Q)$ is defined as

$$d_B(P, Q) = \log \left(\int_{x \in \mathcal{X}: p(x), q(x) > 0} \sqrt{p(x)} \sqrt{q(x)} \, d\mu(x) \right)^{-1}$$

(Bhattacharyya (1943)). Furthermore, for a comparison

$$d_B(\exp, R(0,1)) = \log \left(\int_0^1 \sqrt{e^{-x}} dx \right)^{-1} = -\log(2 - \frac{2}{\sqrt{e}}) \approx 0.239605$$

1.8 Verify $2d_V(P, Q) = \int_{\mathcal{X}} |dP - dQ|$.

1.9 Check the inequalities 1.3.

1.10 Check the inequalities 1.4.

1.11 Compute the Wasserstein-Kantorovich distances $d_W^{(1)}(F, G)$ and $d_W^{(2)}(F, G)$ for the exponential distribution and the uniform distribution (as in Example 1.6).

Chapter 2

Characteristics of robustness

2.1 Influence function

Expansion (1.18) of difference $T(P_n) - T(P)$ says that

$$T(P_n) - T(P) = \frac{1}{n} \sum_{i=1}^{n} T'_{X_i}(P) + n^{-1/2} R_n \qquad (2.1)$$

where the reminder term is asymptotically negligible, $n^{-1/2} R_n = o_p(n^{-1/2})$ as $n \to \infty$. Then we can consider $\frac{1}{n} \sum_{i=1}^{n} T'_{X_i}(P)$ as an error of estimating $T(P)$ by $T(P_n)$, and $T'_{X_i}(P)$ as a contribution of X_i to this error, or as an *influence* of X_i on this error. From this point of view, a natural interpretation of the Gâteau derivative $T'_x(P)$, $x \in \mathcal{X}$ is to call it an *influence function* of functional $T(P)$.

Definition 2.1 *The Gâteau derivative of functional T in distribution P in the direction of Dirac distribution δ_x, $x \in \mathcal{X}$ is called the influence function of T in P; thus*

$$IF(x; T, P) = T'_x(P) = lim_{t \to 0+} \frac{T(P_t(\delta_x)) - T(P)}{t} \qquad (2.2)$$

where $P_t(\delta_x) = (1 - t)P + t\delta_x$.

As the first main properties of IF, let us mention:

a) $\mathbb{E}_P(IF(x; T, P)) = \int_{\mathcal{X}} T'_x(P) dP = 0$,
   hence the average influence of all points x on the estimation error is zero.

b) If T is a Fréchet differentiable functional satisfying condition (1.16), and

$$\mathrm{var}_P(IF(x; T, P)) = \mathbb{E}_P(IF(x; T, P))^2 > 0$$

then $\left(\sqrt{n}(T(P_n) - T(P)) \right) \longrightarrow \mathcal{N}\left(0, \mathrm{var}_P(IF(x; T, P)) \right)$

Example 2.1

(a) *Expected value:* $T(P) = \mathbb{E}_P(X) = m_P$, *then*

$$T(P_n) = \bar{X}_n$$

$$IF(x; T, P) = T'_x(P) = x - m_p$$

$$\mathbb{E}_P(IF(x; T, P)) = 0$$

$$\mathrm{var}_P(IF(x; T, P)) = \mathrm{var}_P X = \sigma_P^2$$

$$\mathbb{E}_Q(IF(x; T, P)) = m_Q - m_P \quad \text{for } Q \neq P$$

$$\mathcal{L}\left(\sqrt{n}(\bar{X}_n - m_p)\right) \longrightarrow \mathcal{N}(0, \sigma_P^2)$$

provided P is the true probability distribution of random sample $(X_1, \ldots, X_n)$.

(b) *Variance:* $T(P) = \mathrm{var}_P X = \sigma_P^2$, *then*

$$IF(x; T, P) = (x - m_P)^2 - \sigma_P^2$$

$$\mathbb{E}_P(IF(x; T, P)) = 0$$

$$\mathrm{var}_P(IF(x; T, P)) = \mu_4 - \mu_2^2 = \mu_4 - \sigma_P^4$$

$$\mathbb{E}_Q(IF(x; T, P)) = \mathbb{E}_Q(X - m_p)^2 - \sigma_P^2$$

$$= \sigma_Q^2 + (m_Q - m_P)^2 + 2\mathbb{E}_Q(X - m_Q)(m_Q - m_P)$$

$$-\sigma_P^2 = \sigma_Q^2 - \sigma_P^2 + (m_Q - m_P)^2$$

2.2   Discretized form of influence function

Let $(X_1, \ldots, X_n)$ be the vector of observations and denote $T_n = T(P_n) = T_n(X_1, \ldots, X_n)$ as its empirical functional. Consider what happens if we add another observation Y to $X_1, \ldots, X_n$. The influence of Y on T_n is characterized by the difference

$$T_{n+1}(X_1, \ldots, X_n, Y) - T_n(X_1, \ldots, X_n) := I(T_n, Y) \qquad (2.3)$$

Because

$$P_n = \frac{1}{n} \sum_{i=1}^{n} \delta_{X_i}$$

$$P_{n+1} = \frac{1}{n+1} \left(\sum_{i=1}^{n} \delta_{X_i} + \delta_Y \right) = \frac{n}{n+1} P_n + \frac{1}{n+1} \delta_Y$$

$$= \left(1 - \frac{1}{n+1} \right) P_n + \frac{1}{n+1} \delta_Y$$

we can say that P_{n+1} arose from P_n by its contamination by the one-point distribution δ_Y in ratio $\frac{1}{n+1}$, hence

$$I(T_n, Y) = T\left[\left(1 - \frac{1}{n+1}\right)P_n + \frac{1}{n+1}\delta_Y\right] - T(P_n)$$

Because

$$\lim_{n \to \infty} (n+1)I(T_n, Y) \tag{2.4}$$

$$= \lim_{n \to \infty} \frac{T\left[\left(1 - \frac{1}{n+1}\right)P_n + \frac{1}{n+1}\delta_Y\right] - T(P_n)}{\frac{1}{n+1}}$$

$$= IF(Y; T, P)$$

$(n+1)I(T_n, Y)$ can be considered as a discretized form of the influence function. The supremum of $|I(T_n, Y)|$ over Y then represents a measure of sensitivity of the empirical functional T_n with respect to an additional observation, under fixed $X_1, \ldots, X_n$.

Definition 2.2 *The number*

$$S(T_n) = \sup_Y |I(T_n(X_1, \ldots, X_n), Y)| \tag{2.5}$$

is called a sensitivity of functional $T_n(X_1, \ldots, X_n)$ to an additional observation.

Example 2.2

(a) Expected value:

$$T(P) = \mathbb{E}_P X, \quad T_n = \bar{X}_n, \quad T_{n+1} = \bar{X}_{n+1}$$

$$\Longrightarrow T_{n+1} = \frac{1}{n+1}(n\bar{X}_n + Y)$$

$$I(T_n, Y) = \left(\frac{n}{n+1} - 1\right)\bar{X}_n + \frac{1}{n+1}Y = \frac{1}{n+1}(Y - \bar{X}_n)$$

$$\Longrightarrow (n+1)I(T_n, Y) = Y - \bar{X}_n \xrightarrow{P} Y - \mathbb{E}_P X \quad as \; n \to \infty$$

$$\Longrightarrow S(\bar{X}_n) = \frac{1}{n+1}\sup_Y |Y - \bar{X}_n| = \infty$$

Thus, the sample mean has an infinite sensitivity to an additional observation.

(b) Median:
Let $n = 2m + 1$ and let $X_{(1)} \leq \cdots \leq X_{(n)}$ be the observations ordered in

increasing magnitude. Then $T_n = T_n(X_1, \ldots, X_n) = X_{(m+1)}$ and $T_{n+1} = T_{n+1}(X_1 \ldots, X_n, Y)$ take on the following values, depending on the position of Y among the other observations:

$$T_{n+1} = \begin{cases} \frac{X_{(m)}+X_{(m+1)}}{2} & \cdots \quad Y \leq X_{(m)} \\ \frac{X_{(m+1)}+X_{(m+2)}}{2} & \cdots \quad Y \geq X_{(m+2)} \\ \frac{Y+X_{(m+1)}}{2} & \cdots \quad X_{(m)} \leq Y \leq X_{(m+2)} \end{cases}$$

Hence, the influence of adding Y to $X_1, \ldots, X_n$ on the median is measured by

$$I(T_n, Y) = \begin{cases} \frac{X_{(m)}-X_{(m+1)}}{2} & Y \leq X_{(m)} \\ \frac{X_{(m+2)}-X_{(m+1)}}{2} & Y \geq X_{(m+2)} \\ \frac{Y-X_{(m+1)}}{2} & X_{(m)} \leq Y \leq X_{(m+2)} \end{cases}$$

Among three possible values of $|I(T_n, Y)|$ is $|\frac{1}{2}(Y - X_{(m+1)})|$ the smallest; thus the sensitivity of the median to an additional observation is equal to

$$S(T_n) = \max\left\{ \frac{1}{2}(X_{(m+1)} - X_{(m)}), \frac{1}{2}(X_{(m+2)} - X_{(m+1)}) \right\}$$

and it is finite under any fixed $X_1, \ldots, X_n$.

2.3   Qualitative robustness

As we have seen in Example 2.1, the influence functions of the expectation and variance are unbounded and can assume arbitrarily large values. Moreover, Example 2.2 shows that adding one more observation can cause a breakdown of the sample mean. The least squares estimator (LSE) behaves analogously (in fact, the mean is a special form of the least squares estimator). Remember the Kagan, Linnik and Rao theorem, mentioned in Section 1.1, that illustrates a large sensitivity of the LSE to deviations from the normal distribution of errors. Intuitively it means that the least squares estimator (and the mean) are very non-robust.

How can we mathematically express this intuitive non-robustness property, and how shall we define the concept of robustness? Historically, this concept has been developing over a rather long period, since many statisticians observed a sensitivity of statistical procedures to deviations from assumed models, and analyzed it from various points of view.

It is interesting that the physicists and astronomers, who tried to determine values of various physical, geophysical and astronomic parameters by means of an average of several measurements, were the first to notice the sensitivity of the mean and the variance to outlying observations. This interesting part of the statistical history is nicely described in the book by Stigler (1986). The history goes up to 1757, when R. J. Boskovich, analyzing his experiments aiming at a characterization of the shape of the globe, proposed an estimation method alternative to the least squares. E. S. Pearson noticed the sensitivity of the classical analysis of variance procedures to deviations from normality in 1931. J. W. Tukey and his Princeton group have started a systematic study of possible alternatives to the least squares since the 1940s. The name "robust" was first used by Box in 1953. Box and Anderson (1955) characterized as robust such a statistical procedure that is not very sensitive to changes of the nuisance or unimportant parameters, while it is sensitive (efficient) to its parameter of interest.

When we speak about robustness of a statistical procedure, we usually mean its robustness with respect to deviations from the assumed distribution of errors. However, other types of robustness are also important, such as the assumed independence of observations, the assumption that is often violated in practice. The first mathematical definition of robustness was formulated by Hampel (1968, 1971), who based the concept of robustness of a statistical functional on its continuity in a neighborhood of the considered probability distribution. The continuity and neighborhood were considered with respect to the Prokhorov metric on the space $\mathcal{P}$.

Let a random variable (or random vector) X take on values in the sample space $(\mathcal{X}, \mathcal{B})$; denote P as its probability distribution. We shall try to characterize mathematically the robustness of the functional $T(P) = T(X)$. This functional is estimated with the aid of observations $X_1, \ldots, X_n$, that are independent copies of X. More precisely, we estimate T by the empirical functional $T_n(P_n) = T_n(X_1, \ldots, X_n)$, based on the empirical distribution P_n of $X_1, \ldots, X_n$. Instead of the empirical functional, T_n is often called the (sample) *statistic*. Hampel's definition of the (qualitative) robustness is based on the Prokhorov metric d_P on the system $\mathcal{P}$ of probability measures on the sample space.

Definition 2.3 *We say that the sequence of statistics (empirical functionals) $\{T_n\}$ is qualitatively robust for probability distribution P, if to any $\varepsilon > 0$ there exists a $\delta > 0$ and a positive integer n_0 such that, for all $Q \in \mathcal{P}$ and $n \geq n_0$,*

$$d_P(P, Q) < \delta \implies d_P(\mathcal{L}_P(T_n), \mathcal{L}_Q(T_n)) < \varepsilon \tag{2.6}$$

where $\mathcal{L}_P(T_n)$ and $\mathcal{L}_Q(T_n)$ denote the probability distributions of T_n under P and Q, respectively.

This robustness is only *qualitative*: it only says whether it is or is not functionally robust, but it does not numerically measure a level of this characteristic. Because such robustness concerns only the behavior of the functional in a small neighborhood of P_0, it is in fact *infinitesimal*. We can obviously replace the Prokhorov metric with another suitable metric on space $\mathcal{P}$, e.g., the Lévy metric.

However, we do not only want to see whether T is or is not robust. We want to compare the various functionals with each other and see which is more robust than the other. To do this, we must characterize the robustness with some quantitative measure. There are many possible quantifications of the robustness. However, using such quantitative measures, be aware that a replacement of a complicated concept with just one number can cause a bias and suppress important information.

2.4   Quantitative characteristics of robustness based on influence function

Influence function is one of the most important characteristics of the statistical functional/estimator. The value $IF(x; T, P)$ measures the effect of a contamination of functional T by a single value x. Hence, a robust functional T should have a bounded influence function. However, even the fact that T is a qualitatively robust functional does not automatically mean that its influence function $IF(x; T, P)$ is bounded. As we see later, an example of such a functional is the R-estimator of the shift parameter, which is an inversion of the van der Waerden rank test; while it is qualitatively robust, its influence function is unbounded.

   The most popular quantitative characteristics of robustness of functional T, based on the influence function, are its *global and local sensitivities:*

a)   *The global sensitivity* of the functional T under distribution P is the maximum absolute value of the influence function in x under P, i.e.,

$$\gamma^* = \sup_{x \in \mathcal{X}} |IF(x; T, P)| \tag{2.7}$$

b)   *The local sensitivity* of the functional T under distribution P is the value

$$\lambda^* = \sup_{x,y;\ x \neq y} \left| \frac{IF(y; T, P) - IF(x; T, P)}{y - x} \right| \tag{2.8}$$

that indicates the effect of the replacement of value x by value y on the functional T.

The following example illustrates the difference between the global and local sensitivities.

Example 2.3

(a) Mean

$T(P) = \mathbb{E}_P(X), \quad IF(x; T, P) = x - \mathbb{E}_P X \implies \gamma^* = \infty, \quad \lambda^* = 1;$
the mean is not robust, but it is not sensitive to the local changes.

(b) Variance

$$T(P) = \mathrm{var}_P X = \sigma_P^2$$

$$IF(x; T, P) = (x - \mathbb{E}_P(X))^2 - \sigma_P^2, \quad \gamma^* = \infty$$

$$\lambda^* = \sup_{y \neq x} \left| \frac{(x - \mathbb{E}_P(X))^2 - (y - \mathbb{E}_P(X))^2}{x - y} \right|$$

$$= \sup_{y \neq x} \left| \frac{x^2 - y^2 - 2(x - y)\mathbb{E}_P X}{x - y} \right| = \sup_{y \neq x} |x + y - 2\mathbb{E}_P X| = \infty$$

hence the variance is non-robust both to large as well as to small (local) changes.

2.5   Maximum bias

Assume that the true distribution function F_0 lies in some family $\mathcal{F}$. Another natural measure of robustness of the functional T is its maximal bias (maxbias) over $\mathcal{F}$,

$$b(\mathcal{F}) = \sup_{F \in \mathcal{F}} \{ |T(F) - T(F_0)| \} \tag{2.9}$$

The family $\mathcal{F}$ can have various forms; for example, it can be a neighborhood of a fixed distribution F_0 with respect to some distance described in Section 1.5. In the robustness analysis, $\mathcal{F}$ is often the ε-contaminated neighborhood of a fixed distribution function F_0, that has the form

$$\mathcal{F}_{F_0,\varepsilon} = \{ F : F = (1 - \varepsilon)F_0 + \varepsilon G, \ G \ \text{unknown distribution function} \} \tag{2.10}$$

The value ε of the contamination ratio is considered as known, and known is also the central distribution function F_0. When estimating the location parameter θ of $F(x - \theta)$, where F is an unknown member of $\mathcal{F}$, then the central distribution F_0 is usually taken as symmetric around zero and unimodal, while the contaminating distribution G can run either over symmetric or asymmetric distribution functions. We then speak about symmetric or asymmetric contaminations.

Many statistical functionals are monotone with respect to the stochastic ordering of distribution functions (or random variables), defined in the following way: Random variable X with distribution function F is stochastically smaller than random variable Y with distribution function G, if

$$F(x) \geq G(x) \quad \forall x \in \mathbb{R}$$

The monotone statistical functional thus attains its maxbias either at the stochastically largest member F_∞ or at the stochastically smallest member $F_{-\infty}$ of $\mathcal{F}_{F_0,\varepsilon}$, hence

$$b(\mathcal{F}_{F_0,\varepsilon}) = \max\{ |T(F_\infty) - T(F_0)|, |T(F_{-\infty}) - T(F_0)| \} \tag{2.11}$$

The following example well illustrates the role of the maximal bias; it shows that while the mean is non-robust, the median is universally robust with respect to the maxbias criterion.

Example 2.4

(i) Mean

$T(F) = \mathbb{E}_F(X)$; *if F_0 is symmetric around zero and so are all contaminating distributions G, all having finite first moments, then $T(F)$ is unbiased for all $F \in \mathcal{F}_{F_0,\varepsilon}$, hence $b(\mathcal{F}_{F_0,\varepsilon}) = 0$. However, under an asymmetric contamination, $b(\mathcal{F}_{F_0,\varepsilon}) = |\mathbb{E}(F_\infty) - \mathbb{E}(F_0)| = \infty$, where $F_\infty = (1 - \varepsilon)F_0 + \varepsilon\delta_\infty$, the stochastically largest member of $\mathcal{F}_{F_0,\varepsilon}$.*

(ii) Median

Because the median is nondecreasing with respect to the stochastic ordering of distributions, its maximum absolute bias over an asymmetric ε-contaminated neighborhood of a symmetric distribution function F_0 is attained either at $F_\infty = (1 - \varepsilon)F_0 + \varepsilon\delta_\infty$ (the stochastically largest distribution of $\mathcal{F}_{F_0,\varepsilon}$), or at $F_{-\infty} = (1 - \varepsilon)F_0 + \varepsilon\delta_{-\infty}$ (the stochastically smallest distribution of $\mathcal{F}_{F_0,\varepsilon}$). The median of F_∞ is attained at x_0 satisfying

$$(1 - \varepsilon)F_0(x_0) = \frac{1}{2} \implies x_0 = F_0^{-1}\left(\frac{1}{2(1 - \varepsilon)}\right)$$

while the median of $F_{-\infty}$ is x_0^- such that

$$(1 - \varepsilon)F_0(x_0^-) + \varepsilon = \frac{1}{2} \implies x_0^- = F_0^{-1}\left(1 - \frac{1}{2(1 - \varepsilon)}\right) = -x_0$$

hence the maxbias of the median is equal to x_0.

Let $T(F)$ be any other functional such that its estimate $T(F_n) = T(X_1, \dots, X_n)$, based on the empirical distribution function F_n, is translation equivariant, i.e., $T(X_1 + c, \dots, X_n + c) = T(X_1, \dots, X_n) + c$ for any $c \in \mathbb{R}$. Then obviously $T(F(\cdot - c)) = T(F(\cdot)) + c$. We shall show that the maxbias of T cannot be smaller than x_0.

Consider two contaminations of F_0,

$$F_+ = (1 - \varepsilon)F_0 + \varepsilon G_+, \quad F_- = (1 - \varepsilon)F_0 + \varepsilon G_-$$

where

$$G_+(x) = \begin{cases} 0 & \cdots \quad x \leq x_0 \\ \frac{1}{\varepsilon}\{1 - (1 - \varepsilon)[F_0(x) + F_0(2x_0 - x)]\} & \cdots \quad x \geq x_0 \end{cases}$$

and

$$G_-(x) = \begin{cases} \frac{1}{\varepsilon}\{(1 - \varepsilon)[F_0(x + 2x_0) - F_0(x)]\} & \cdots \quad x < -x_0 \\ 1 & \cdots \quad x \geq -x_0 \end{cases}$$

Notice that $F_-(x - x_0)) = F_+(x + x_0)$, hence $T(F_-) + x_0 = T(F_+) - x_0$ and $T(F_+) - T(F_-) = 2x_0$; thus the maxbias of T at F_0 cannot be smaller than x_0. It shows that the median has the smallest maxbias among all translation equivariant functionals.

If $T(F)$ is a nonlinear functional, or if it is defined implicitly as a solution of a minimization or of a system of equations; then it is difficult to calculate (2.11) precisely. Then we consider the *maximum asymptotic bias* of $T(F)$ over a neighborhood $\mathcal{F}$ of F_0. More precisely, let $X_1, \ldots, X_n$ be independent identically distributed observations, distributed according to distribution function $F \in \mathcal{F}$ and F_n be the empirical distribution function. Assume that under an infinitely increasing number of observations, $T(F_n)$ has an asymptotical normal distribution for every $F \in \mathcal{F}$ in the sense that

$$P\left(\sqrt{n}(T(F_n) - T(F)) \leq x\right) \to \Phi\left(\frac{x}{\sigma(T, F)}\right) \quad \text{as} \ \ n \to \infty$$

with variance $\sigma^2(T, F)$ dependent on T and F. Then the maximal asymptotic bias (asymptotic maxbias) of T over $\mathcal{F}$ is defined as

$$\sup\left\{|T(F) - T(F_0)| : F \in \mathcal{F}\right\} \tag{2.12}$$

We shall return to the asymptotic maxbias later in the context of some robust estimators that are either nonlinear or defined implicitly as a solution of a minimization or a system of equations.

2.6   Breakdown point

The breakdown point, introduced by Donoho and Huber in 1983, is a very popular quantitative characteristic of robustness. To describe this characteristic, start from a random sample $\boldsymbol{x}^0 = (x_1, \ldots, x_n)$ and consider the corresponding value $T_n(\boldsymbol{x}^0)$ of an estimator of functional T. Imagine that in this "initial" sample we can replace any m components by arbitrary values, possibly very unfavorable, even infinite. The new sample after the replacement denotes $\boldsymbol{x}^{(m)}$, and let $T_n(\boldsymbol{x}^m)$ be the pertaining value of the estimator.

The *breakdown point* of estimator T_n for sample $\boldsymbol{x}^{(0)}$ is the number

$$\varepsilon_n^*(T_n, \boldsymbol{x}^{(0)}) = \frac{m^*(\boldsymbol{x}^{(0)})}{n}$$

where $m^*(\boldsymbol{x}^{(0)})$ is the smallest integer m, for which

$$\sup_{\boldsymbol{x}^{(m)}} \|T_n(\boldsymbol{x}^{(m)}) - T_n(\mathbf{x}^{(0)})\| = \infty$$

i.e., the smallest part of the observations that, being replaced with arbitrary values, can lead T_n up to infinity. Some estimators have a universal breakdown point, when m^* is independent of the initial sample $\boldsymbol{x}^{(0)}$. Then we can also calculate the limit $\varepsilon^* = \lim_{n\to\infty} \varepsilon_n^*$, which is often also called the breakdown point.

We can modify the breakdown point in such a way that, instead of replacing m components, we extend the sample by some m (unfavorable) values.

Example 2.5

(a) The average $\bar{X}_n = \frac{1}{n}\sum_{i=1}^{n} X_i$:

$\varepsilon_n^*(\bar{X}_n, \boldsymbol{x}^{(0)}) = \frac{1}{n}$, *hence* $\lim_{n\to} \varepsilon_n^*(\bar{X}_n, \boldsymbol{x}^{(0)}) = 0$ *for any initial sample* $\boldsymbol{x}^{(0)}$

(b) Median $\tilde{X}_n = X_{\left(\frac{n+1}{2}\right)}$ (consider n odd, for simplicity):

$\varepsilon_n^*(\tilde{X}_n, \boldsymbol{x}^{(0)}) = \frac{n+1}{2n}$, *thus* $\lim_{n\to} \varepsilon_n^*(\tilde{X}_n, \boldsymbol{x}^{(0)}) = \frac{1}{2}$ *for any initial sample* $\boldsymbol{x}^{(0)}$

2.7　Tail–behavior measure of a statistical estimator

The tail–behavior measure is surprisingly intuitive mainly in estimating the shift and regression parameters. We will first illustrate this measure on the shift parameter, and then return to regression at a suitable place.

Let $(X_1, \ldots, X_n)$ be a random sample from a population with continuous distribution function $F(x - \theta)$, $\theta \in \mathbb{R}$. The problem of interest is that of estimating parameter θ. A reasonable estimator of the shift parameter should be *translation equivariant*: T_n is translation equivariant, if

$$T_n(X_1 + c, \ldots, X_n + c) = T_n(X_1, \ldots, X_n) + c$$

$$\forall\, c \in \mathbb{R} \text{ and } \forall X_1 \ldots, X_n$$

The performance of such an estimator can be characterized by probabilities

$$P_\theta(|T_n - \theta| > a)$$

analyzed either under fixed $a > 0$ and $n \to \infty$, or under fixed n and $a \to \infty$. Indeed, if $\{T_n\}$ is a consistent estimator of θ, then $\lim_{n\to 0} P_\theta(|T_n - \theta| > a) = 0$ under any fixed $a > 0$. Such a characteristic was studied, e.g., by Bahadur (1967), Fu (1975, 1980) and Sievers (1978), who suggested the limit

$$\lim_{n\to\infty} \left\{ -\frac{1}{n}\ln P_\theta(|T_n - \theta| > a) \right\} \quad \text{under fixed } a > 0$$

(provided that it exists) as a measure of efficiency of estimator T_n, and compared estimators from this point of view.

On the other hand, a good estimator T_n also verifies the convergence

$$\lim_{a\to\infty} P_\theta(|T_n - \theta| > a) = \lim_{a\to\infty} P_0(|T_n| > a) = 0 \qquad (2.13)$$

while this convergence is as fast as possible. The probabilities $P_\theta(T_n - \theta > a)$ and $P_\theta(T_n - \theta < -a)$, for a sufficiently large, are called the right and the left tails, respectively, of the probability distribution of T_n. If T_n is symmetrically distributed around θ, then both its tails are characterized by probability (2.13). This probability should rapidly tend to zero. However, the speed of this convergence cannot be arbitrarily high. We shall show that the rate of convergence of tails of a translation equivariant estimator is bounded, and that its upper bound depends on the behavior of $1 - F(a)$ and $F(-a)$ for large $a > 0$.

Let us illustrate this upper bound on a model with symmetric distribution function satisfying $F(-x) = 1 - F(x)$ $\forall x \in \mathbb{R}$. Jurečková (1981) introduced the following tail-behavior measure of an equivariant estimator T_n :

$$B(T_n; a) = \frac{-\ln P_\theta(|T_n - \theta| > a)}{-\ln(1 - F(a))} = \frac{-\ln P_0(|T_n| > a)}{-\ln(1 - F(a))}, \quad a > 0 \qquad (2.14)$$

The values $B(T_n; a)$ for $a > 0$ show how many times faster the probability $P_0(|T_n| > a)$ tends to 0 than $1 - F(a)$, as $a \to \infty$. The best is estimator T_n with the largest possible values $B(T_n; a)$ for $a > 0$. The lower and upper bounds for $B(T_n; a)$, thus for the rate of convergence of its tails, are formulated in the following lemma:

Lemma 2.1 *Let $X_1, \ldots, X_n$ be a random sample from a population with distribution function $F(x - \theta)$, $0 < F(x) < 1$, $F(-x) = 1 - F(x)$, $x, \theta \in \mathbb{R}$. Let T_n be an equivariant estimator of θ such that, for any fixed n*

$$\min_{1 \leq i \leq n} X_i > 0 \implies T_n(X_1, \ldots, X_n) > 0$$

$$(2.15)$$

$$\max_{1 \leq i \leq n} X_i < 0 \implies T_n(X_1, \ldots, X_n) < 0$$

Then, under any fixed n

$$1 \leq \underline{\lim}_{a \to \infty} B(T_n; a) \leq \overline{\lim}_{a \to \infty} B(T_n; a) \leq n \qquad (2.16)$$

Proof. Indeed, if T_n is equivariant, then

$$P_0(|T_n(X_1, \ldots, X_n)| > a)$$

$$= P_0(T_n(X_1, \ldots, X_n) > a) + P_0(T_n(X_1, \ldots, X_n) < -a)$$

$$= P_0(T_n(X_1 - a, \ldots, X_n - a) > 0) + P_0(T_n(X_1 + a, \ldots, X_n + a) < 0)$$

$$\geq P_0\left(\min_{1 \leq i \leq n} X_i > a\right) + P_0\left(\max_{1 \leq i \leq n} X_i < -a\right)$$

$$= (1 - F(a))^n + (F(-a))^n$$

hence

$$-\ln P_0(|T_n(X_1, \ldots, X_n)| > a) \leq -\ln 2 - n \ln(1 - F(a))$$

$$\implies \overline{\lim}_{a \to \infty} \frac{-\ln P_0(|T_n| > a)}{-\ln(1 - F(a))} \leq n$$

Similarly,

$$P_0(|T_n(X_1,\ldots,X_n)| > a)$$

$$\leq P_0\left(\min_{1\leq i\leq n} X_i \leq -a\right) + P_0\left(\max_{1\leq i\leq n} X_i \geq a\right)$$

$$= 1 - (1 - F(-a))^n + 1 - (F(a))^n = 2\{1 - (F(a))^n\}$$

$$= 2(1 - F(a))\left[1 + F(a) + \ldots + (F(a))^{n-1}\right]$$

$$\leq 2n(1 - F(a))$$

hence

$$-\ln P_0(|T_n(X_1,\ldots,X_n)| > a) \geq -\ln 2 - \ln n - \ln(1 - F(a))$$

$$\implies \underline{\lim}_{a\to\infty} \frac{-\ln P_0(|T_n| > a)}{-\ln(1 - F(a))} \geq 1$$

$\square$

If T_n attains the upper bound in (2.16), then it is obviously optimal for distribution function F, because its tails tend to zero n-times faster than $1 - F(a)$, which is the upper bound. However, we still have the following questions:

- Is the upper bound attainable, and for which T_n and F?
- Is there any estimator T_n attaining high values of $B(T_n; a)$ robustly for a broad class of distribution functions?

It turns out that the sample mean $\bar{X}_n$ can attain both lower and upper bounds in (2.16); namely, it attains the upper bound under the normal distribution and under an exponentially tailed distribution, while it attains the lower bound only for the Cauchy distribution and for the heavy-tailed distributions. This demonstrates a high non-robustness of $\bar{X}_n$ even from the tail behavior aspect. On the other hand, the sample median $\tilde{X}_n$ is robust even with respect to tails: $\tilde{X}_n$ does not attain the upper bound in (2.16), on the contrary, the limit $\lim_{a\to\infty} B(\tilde{X}_n; a)$ is always in the middle of the scope between 1 and n for a broad class of distribution functions.

These conclusions are in good concordance with the robustness concepts. The following theorem gives them a mathematical form.

Theorem 2.1 *Let $X_1,\ldots,X_n$ be a random sample from a population with distribution function $F(x - \theta)$, $0 < F(x) < 1$, $F(-x) = 1 - F(x)$, $x, \theta \in \mathbb{R}$.*

(i)   Let $\bar{X}_n = \frac{1}{n}\sum_{i=1}^{n} X_i$ be the sample mean. If F has exponential tails, i.e.,

$$\lim_{a\to\infty} \frac{-\ln(1 - F(a))}{ba^r} = 1 \quad \text{for some } b > 0, \ r \geq 1 \qquad (2.17)$$

then

$$\lim_{a \to \infty} B(\bar{X}_n; a) = n \tag{2.18}$$

(ii) If F has heavy tails in the sense that

$$\lim_{a \to \infty} \frac{-\ln(1 - F(a))}{m \ln a} = 1 \quad \text{for some } m > 0 \tag{2.19}$$

then

$$\lim_{a \to \infty} B(\bar{X}_n; a) = 1 \tag{2.20}$$

(iii) Let $\tilde{X}_n$ be the sample median. Then for F satisfying either (2.17) or (2.19),

$$\frac{n}{2} \leq \overline{\lim}_{a \to \infty} B(\tilde{X}_n; a) \leq \frac{n}{2} + 1 \quad \text{for } n \text{ even, and} \tag{2.21}$$

$$\lim_{a \to \infty} B(\tilde{X}_n, a) = \frac{n+1}{2} \quad \text{for } n \text{ odd} \tag{2.22}$$

Remark 2.1 *The distribution functions with exponential tails, satisfying (2.17), will be briefly called type I. This class includes the normal distribution ($r = 2$), logistic and the Laplace distributions ($r = 1$). The distribution functions with heavy tails, satisfying (2.19), will be called type II. The Cauchy distribution ($m = 1$) or the t-distribution with $m > 1$ degrees of freedom belongs here.*

Proof of Theorem 2.1.
(i) It is sufficient to prove that the exponentially tailed F has a finite expected value

$$E_\varepsilon = \mathbb{E}_0 \left[\exp \left\{ n(1 - \varepsilon) b |\bar{X}_n|^r \right\} \right] < \infty \tag{2.23}$$

for arbitrary $\varepsilon \in (0, 1)$. Indeed, then we conclude from the Markov inequality that

$$P_0(|\bar{X}_n| > a) \leq E_\varepsilon \cdot \exp\{-n(1 - \varepsilon) b a^r\}$$

$$\implies \underline{\lim}_{a \to \infty} \frac{-\ln P_0(|\bar{X}_n| > a)}{b a^r} \geq \lim_{a \to \infty} \frac{n(1 - \varepsilon) b a^r - \ln E_\varepsilon}{b a^r} = n(1 - \varepsilon)$$

and we arrive at proposition (2.18).
    The finite expectation (2.23) we get from the Hölder inequality:

$$\mathbb{E}_0 \left[\exp \left\{ n(1 - \varepsilon) b |\bar{X}_n|^r \right\} \right] \leq \mathbb{E}_0 \left[\exp \left\{ (1 - \varepsilon) b \sum_{i=1}^n |X_i|^r \right\} \right] \tag{2.24}$$

$$\leq (\mathbb{E}_0 \left[\exp \left\{ (1 - \varepsilon) b |X_1|^r \right\} \right])^n = 2^n \left(\int_0^\infty \left[\exp \left\{ (1 - \varepsilon) b x^r \right\} \right] dF(x) \right)^n$$

It follows from (2.17) that, given $\varepsilon > 0$, there exists an $A_\varepsilon > 0$ such that

$$1 - F(a) < \exp\left\{-(1 - \frac{\varepsilon}{2})ba^r\right\}$$

holds for any $a \geq A_\varepsilon$.

The last integral in (2.24) can be successively rewritten in the following way:

$$\int_0^\infty \exp\left\{(1 - \varepsilon)bx^r\right\} dF(x) = \int_0^{A_\varepsilon} \exp\left\{(1 - \varepsilon)bx^r\right\} dF(x)$$

$$- \int_{A_\varepsilon}^\infty \exp\left\{(1 - \varepsilon)bx^r\right\} d(1 - F(x))$$

$$= \int_0^{A_\varepsilon} \exp\left\{(1 - \varepsilon)bx^r\right\} dF(x) + (1 - F(A_\varepsilon)) \cdot \exp\left\{(1 - \varepsilon)bA_\varepsilon^r\right\}$$

$$+ \int_{A_\varepsilon}^\infty (1 - F(x))(1 - \varepsilon)brx^{r-1} \cdot \exp\left\{(1 - \varepsilon)bx^r\right\} dx$$

$$\leq \int_0^{A_\varepsilon} \exp\left\{(1 - \varepsilon)bx^r\right\} dF(x) + \exp\left\{-\frac{\varepsilon}{2}bA_\varepsilon^r\right\}$$

$$+ \int_{A_\varepsilon}^\infty (1 - \varepsilon)brx^{r-1} \cdot \exp\left\{-\frac{\varepsilon}{2}bx^r\right\} dx < \infty$$

and that leads to proposition (i).

(ii) If F has heavy tails, then

$$P_0(|\bar{X}_n| > a) = P_0(\bar{X}_n > a) + P_0(\bar{X}_n < -a)$$

$$\geq P_0\left(X_1 > -a, \ldots, X_{n-1} > -a, X_n > (2n - 1)a\right)$$

$$+ P_0\left(X_1 < a, \ldots, X_{n-1} < a, X_n < -(2n - 1)a\right)$$

$$= 2(F(a))^{n-1}[1 - F((2n - 1)a)]$$

hence

$$\overline{\lim}_{a \to \infty} B(\bar{X}_n, a) \leq \overline{\lim}_{a \to \infty} \frac{-\ln\left[1 - F(2n - 1)a\right]}{m \ln a}$$

$$= \lim_{a \to \infty} \frac{-\ln\left[1 - F(2n - 1)a\right]}{m \ln((2n - 1)a)} = 1$$

(iii) Let $\tilde{X}_n$ be the sample median and n be odd. Then $\tilde{X}_n$ is the middle-order statistic of the sample $X_1, \ldots, X_n$, i.e., $\tilde{X}_n = X_{(m)}$, $m = \frac{n+1}{2}$, and $F(\tilde{X}_n) = U_{(m)}$ has the beta-distribution. Then

$$P_0(|\tilde{X}_n| > a) = P_0(\tilde{X}_n > a) + P_0(\tilde{X}_n < -a)$$

$$= 2n \binom{n-1}{m-1} \int_{F(a)}^1 u^{m-1}(1-u)^{m-1} du$$

$$\leq 2n \binom{n-1}{m-1} (1 - F(a))^m$$

and similarly

$$P_0(|\tilde{X}_n| > a) \geq 2n \binom{n-1}{m-1} (F(a))^{m-1}(1 - F(a))^m$$

that leads to (2.22) after taking logarithms. Analogously we proceed for n even. □

2.8 Variance of asymptotic normal distribution

If estimator T_n of functional $T(\cdot)$ is asymptotically normally distributed as $n \to \infty$,

$$\mathcal{L}_P\left(\sqrt{n}(T_n - T(P))\right) \to \mathcal{N}(0, V^2(P,T))$$

then another possible robustness measure of T is the supremum of the variance $V^2(P,T)$

$$\sigma^2(T) = \sup_{P \in \mathcal{P}_0} V^2(P,T)$$

over a neighborhood $\mathcal{P}_0 \subset \mathcal{P}$ of the assumed model.

The estimator minimizing $\sup_{P \in \mathcal{P}_0} V^2(P,T)$ over a specified class $\mathcal{T}$ of estimators of parameter θ, is called *minimaximally robust* in the class $\mathcal{T}$. We shall show in the sequel that the classes of M-estimators, L-estimators and R-estimators all contain a minimaximally robust estimator of the shift and regression parameters in a class of contaminated normal distributions.

2.9 Available "robust" packages in R

There are plenty of packages for robust statistical methods available on CRAN. See the overview of the "robust" packages at https://cran.r-project.org/view=Robust. However, here we list only those packages that are used in this book.

stats is the base R package that contains the basic robust statistical functions like median, trimmed mean, median absolute deviation or inter-quartile range.

MASS is a recommended package, part of the standard R distribution. It was created for the book by Venables and Ripley (2002). Its functions `huber`, `hubers`, `rlm` and `lqs` can be used for computation of the M-estimator, GM-estimator, MM-estimator, least median of squares estimator or least trimmed squares estimator in a location or a linear model.

robustbase is an R package for basic robust statistics and covers the book by Maronna, Martin and Yohai (2006). Its function `lmrob` is used here to compute the M-estimator, S-estimator and MM-estimator.

lmomco is a package that comprehensively implements L-moments, Censored L-moments, Trimmed L-moments, L-Comoments, and Many Distributions. Some L-estimators like the Gini mean difference or Sen's weighted mean can be computed by the functions from this package.

quantreg is a package for quantile regression. It includes estimation and inference methods for linear and nonlinear parametric and non-parametric models. The function `rq` can be employed to estimate the regression quantile as well as the regression rank scores process in linear model. Also the test based on regression rank scores is implemented.

Rfit is a package for rank-based estimation and inference. Function `rfit` is used to minimize the Jaeckel's dispersion function.

mblm provides estimation for a linear model with high breakdown point. Here it is used to obtain the repeated median in a simple regression model.

galts is a package that includes function `ga.lts` that estimates the parameters of least trimmed squares estimator using genetic algorithms.

2.10 Problems and complements

2.1 Show that both the sample mean and the sample median of the random sample $X_1, \ldots, X_n$ are nondecreasing in each argument X_i, $i = 1, \ldots, n$.

2.2 Characterize distributions satisfying

$$\lim_{a \to \infty} \frac{-\ln(1 - F(a + c))}{-\ln(1 - F(a))} = 1 \tag{2.25}$$

for any fixed $c \in \mathbb{R}$. Show that this class contains distributions of Type 1 and Type 2.

2.3 Let $X_1, \ldots, X_n$ be a random sample from a population with distribution function $F(x - \theta)$, where F is symmetric, absolutely continuous, $0 < F(x) < 1$ for $x \in \mathbb{R}$, and satisfying (2.25). Let $T_n(X_1, \ldots, X_n)$ be a translation equivariant estimator of θ, nondecreasing in each argument X_i, $i = 1, \ldots, n$. Then T_n has a universal breakdown point $m_n^* = m_n^*(T_n)$ and there exists a constant A such that

$$X_{n:m_n^*} - A \leq T_n(X_1, \ldots, X_n) \leq X_{n:n-m_n^*+1} + A$$

where $X_{n:1} \leq X_{n:2} \leq \ldots \leq X_{n:n}$ are the order statistics of the sample $X_1, \ldots, X_n$. (Hint: see He at al. (1990).)

2.4 Let $T_n(X_1, \ldots, X_n)$ be a translation equivariant estimator of θ, nondecreasing in each argument X_i, $i = 1, \ldots, n$. Then, under the conditions of Problem 2.2,

$$m_n^* \leq \underline{\lim}_{a \to \infty} B(a, T_n) \leq \overline{\lim}_{a \to \infty} B(a, T_n) \leq n - m_n^* + 1 \qquad (2.26)$$

Illustrate it on the sample median. (Hint: see He et al. (1990).)

2.5 Let $T_n(X_1, \ldots, X_n)$ be a random sample from a population with distribution function $F(x - \theta)$. Compute the breakdown point of

$$T_n = \frac{1}{2}(X_{n:1} + X_{n:n})$$

This estimator is called the midrange (see the next chapter).

2.6 Show that the midrange (Problem 2.5) of the random sample $X_1, \ldots, X_n$ is nondecreasing in each argument X_i, $i = 1, \ldots, n$. Illustrate (2.26) for this estimator.

2.7 Determine whether the gamma distribution (Example 1.5) has exponential or heavy tails.

Chapter 3

Estimation of real parameter

Let $X_1, \ldots, X_n$ be a random sample from a population with probability distribution P; the distribution is generally unknown, we only assume that its distribution function F belongs to some class $\mathcal{F}$ of distribution functions. We look for an appropriate estimator of parameter θ, that can be expressed as a functional $T(P)$ of P. The same parameter θ can be characterized by means of more functionals, e.g., the center of symmetry is simultaneously the expected value, the median, the modus of the distribution, and other possible characterizations. Some functionals $T(P)$ are characterized implicitly as a root of an equation (or of a system of equations) or as a solution of a minimization (maximization) problem: such are the maximal likelihood estimator, moment estimator, etc. An estimator of parameter θ is obtained as an empirical functional, i.e., when one replaces P in the functional $T(\cdot)$ with the empirical distribution corresponding to the vector of observations $X_1, \ldots, X_n$.

We shall mainly deal with three broad classes of robust estimators of the real parameter: M-estimators, L-estimators, and R-estimators, which we shall briefly supplement with more up-to-date S-, MM- and τ-estimators. The next chapter extends these classes to the linear regression model.

3.1 M-estimators

The class of M-estimators was introduced by P. J. Huber in (1964); the properties of M-estimators are studied in his book (Huber (1981)), and also in the books by Andrews et al. (1972), Antoch et al. (1998), Bunke and Bunke (1986), Dodge and Jurečková (2000), Hampel et al. (1986), Jurečková and Sen (1996), Lecoutre and Tassi (1987), Rieder (1994), Rousseeuw and Leroy (1987), Staudte and Sheather (1990), and others.

M-estimator T_n is defined as a solution of the minimization problem

$$\sum_{i=1}^{n} \rho(X_i, \theta) := \min \quad \text{with respect to} \ \ \theta \in \Theta$$

or   (3.1)

$$\mathbb{E}_{P_n} \left[\rho(X, \theta) \right] = \min, \quad \theta \in \Theta$$

where $\rho(\cdot, \cdot)$ is a properly chosen function. The class of M-estimators covers also the *maximal likelihood estimator* (MLE) of parameter θ in the parametric model $\mathcal{P} = \{P_\theta,\ \theta \in \boldsymbol{\Theta}\}$; if $f(x, \theta)$, is the density function of P_θ, then the MLE is a solution of the minimization

$$\sum_{i=1}^{n} (-\log\ f(X_i, \theta)) = \min, \quad \theta \in \boldsymbol{\Theta}$$

If ρ in (3.1) is differentiable in θ with a continuous derivative $\psi(\cdot, \theta) = \frac{\partial}{\partial \theta}\rho(\cdot, \theta)$, then T_n is a root (or one of the roots) of the equation

$$\sum_{i=1}^{n} \psi(X_i, \theta) = 0, \quad \theta \in \boldsymbol{\Theta} \tag{3.2}$$

hence

$$\frac{1}{n}\sum_{i=1}^{n} \psi(X_i, T_n) = \mathbb{E}_{P_n}\left[\psi(X, T_n)\right] = 0 \quad T_n \in \boldsymbol{\Theta}. \tag{3.3}$$

We see from (3.1) and (3.3) that the M-functional, the statistical functional corresponding to T_n, is defined as a solution of the minimization

$$\int_{\mathcal{X}} \rho(x, T(P))\ dP(x) = \mathbb{E}_P\left[\rho(X, T(P))\right] := \min, \quad T(P) \in \boldsymbol{\Theta} \tag{3.4}$$

or as a solution of the equation

$$\int_{\mathcal{X}} \psi(x, T(P))\ dP(x) = \mathbb{E}_P\left[\psi(X, T(P))\right] = 0, \quad T(P) \in \boldsymbol{\Theta} \tag{3.5}$$

The functional $T(P)$ is Fisher consistent, if the solutions of (3.4) and (3.5) are uniquely determined.

3.1.1   Influence function of M-estimator

Assume that $\rho(\cdot, \theta)$ is differentiable, that its derivative $\psi(\cdot, \theta)$ is absolutely continuous with respect to θ, and that the Equation (3.5) has a unique solution $T(P)$. Let $P_t = (1 - t)P + t\delta_x$; then $T(P_t)$ solves the equation

$$\int_{\mathcal{X}} \psi(y, T(P_t))d((1 - t)P + t\delta_x) = 0$$

hence

$$(1 - t)\int_{\mathcal{X}} \psi(y, T(P_t))\ dP(y) + t\psi(x, T(P_t)) = 0 \tag{3.6}$$

Differentiating (3.6) in t, we obtain

$$-\int_{\mathcal{X}} \psi(y, T(P_t))dP(y) + \psi(x, T(P_t))$$

$$+(1 - t)\frac{dT(P_t)}{dt}\int_{\mathcal{X}}\left[\frac{\partial}{\partial \theta}\psi(y, \theta)\right]_{\theta = T(P_t)} dP(y)$$

$$+t\frac{dT(P_t)}{dt}\left[\frac{\partial}{\partial \theta}\psi(x, \theta)\right]_{\theta = T(P_t)} = 0$$

We obtain the influence function of $T(P)$ if $t \downarrow 0$:

$$IF(x; T, P) = \frac{\psi(x, T(P))}{-\int_{\mathcal{X}} \dot{\psi}(y, T(P) dP(y)}$$ (3.7)

where $\dot{\psi}(y, T(P)) = \left[\frac{\partial}{\partial \theta} \psi(y, \theta)\right]_{\theta = T(P)}$

3.2 *M*-estimator of location

An important special case is the model with the shift parameter θ, where $X_1, \ldots, X_n$ are independent observations with the same distribution function $F(x - \theta)$, $\theta \in \mathbb{R}$; the distribution function F is generally unknown. *M*-estimator T_n is defined as a solution of the minimization

$$\sum_{i=1}^{n} \rho(X_i - \theta) := \min$$ (3.8)

and if $\rho(\cdot)$ is differentiable with absolutely continuous derivative $\psi(\cdot)$, then T_n solves the equation

$$\sum_{i=1}^{n} \psi(X_i - \theta) = 0$$ (3.9)

The corresponding *M*-functional $T(F)$ is Fisher consistent, provided the minimization $\int_{\mathcal{X}} \rho(x - \theta) dP(x) := \min$ has a unique solution $\theta = 0$. The influence function of $T(F)$ is then

$$IF(x; T, P) = \frac{\psi(x - T(P))}{\int_{\mathcal{X}} \psi'(y) dP(y)}$$ (3.10)

We see from the minimization (3.8) and from Equation (3.9) that T_n is *translation equivariant*, i.e., that it satisfies

$$T_n(X_1 + c, \ldots, X_n + c) = T_n(X_1, \ldots, X_n) + c \quad \forall c \in \mathbb{R}$$ (3.11)

However, T_n generally *is not scale equivariant*: the scale equivariance of T_n means that

$$T_n(cX_1, \ldots, cX_n) = cT_n(X_1 \ldots, X_n) \quad \text{for } c > 0$$

If the model is symmetric, i.e., we have a reason to assume the symmetry of F around 0, we should choose ρ symmetric around 0 (ψ would be then an odd function). If $\rho(x)$ is strictly convex (and thus $\psi(x)$ strictly increasing), then $\sum_{i=1}^{n} \rho(X_i - \theta)$ is strictly convex in θ, and the *M*-estimator is uniquely determined. If $\rho(x)$ is linear in some interval $[a, b]$, then $\psi(\cdot)$ is constant in $[a, b]$, and the equation $\sum_{i=1}^{n} \psi(X_i - \theta) = 0$ can have more roots. There are many possible rules for choosing one among these roots; one possibility to obtain a unique solution is to define T_n in the following way:

$$T_n = \frac{1}{2}(T_n^+ + T_n^-)$$

$$T_n^- = \sup\{t : \sum_{i=1}^n \psi(X_i - t) > 0\} \tag{3.12}$$

$$T_n^+ = \inf\{t : \sum_{i=1}^n \psi(X_i - t) < 0\}$$

Similarly, we determine the M-estimator in the case of nondecreasing ψ with jump discontinuities. If $\psi(\cdot)$ is nondecreasing, continuous or having jump discontinuities; then the M-estimator T_n obviously satisfies for any $a \in \mathbb{R}$

$$P_\theta\Big(\sum_{i=1}^n \psi(X_i - a) > 0\Big) \le P_\theta(T_n > a) \le P_\theta(T_n \ge a)$$

$$\le P_\theta\Big(\sum_{i=1}^n \psi(X_i - a) \ge 0\Big) \tag{3.13}$$

$$= P_\theta\Big(\sum_{i=1}^n \psi(X_i - a) > 0\Big) + P_\theta\Big(\sum_{i=1}^n \psi(X_i - a) = 0\Big)$$

The inequalities in (3.13) convert to equalities if $P_\theta\{\sum_{i=1}^n \psi(X_i - a) = 0\} = 0$. This further implies that, for any $y \in \mathbb{R}$,

$$P_0\Big\{n^{-\frac{1}{2}}\sum_{i=1}^n \psi\Big(X_i - \frac{y}{\sqrt{n}}\Big) < 0\Big\} \le P_\theta(\sqrt{n}(T_n - \theta) < y)$$

$$\tag{3.14}$$

$$\le P_\theta(\sqrt{n}(T_n - \theta) \le y) \le P_0\Big\{n^{-\frac{1}{2}}\sum_{i=1}^n \psi\Big(X_i - \frac{y}{\sqrt{n}}\Big) \le 0\Big\}$$

Because $\frac{1}{\sqrt{n}}\sum_{i=1}^n \psi\Big(X_i - \frac{y}{\sqrt{n}}\Big)$ is a standardized sum of independent identically distributed random variables, the asymptotic probability distribution of $\sqrt{n}(T_n - \theta)$ can be derived from the central limit theorem by means of (3.14).

3.2.1　Breakdown point of M-estimator of location parameter

If M_n estimates the center of symmetry θ of $F(x - \theta)$, then its breakdown point follows from Section 2.6:

$\varepsilon^* = \lim_{n\to\infty} \varepsilon_n^* = 0$　　if ψ is an unbounded function

$\varepsilon^* = \lim_{n\to\infty} \varepsilon_n^* = \frac{1}{2}$　　if ψ is odd and bounded

Hence, the class of *M*-estimators contains robust as well as non-robust elements.

Example 3.1

(a) Expected value:

Expected value $\theta = \mathbb{E}_P X$ *is an M-functional with the criterion function* $\rho(x) = x^2$, $\psi(x) = 2x$ *and* $\psi'(x) = 2$. *Its influence function follows from (3.10):*

$$IF(x; T, P) = \frac{2(x - \mathbb{E}_P(X))}{\int_{\mathbf{R}} 2dP} = x - \mathbb{E}_P(X)$$

The corresponding M-estimator is the sample (arithmetic) mean $\bar{X}_n$; *its break-down point* $\varepsilon^* = \lim_{n \to \infty} \varepsilon_n^*$ *is equal to 0, and its global sensitivity* $\gamma^* = +\infty$. *There is a function* mean *in R for the computation of the sample mean.*

(b) Median:

Median $\tilde{X} = F^{-1}(\frac{1}{2})$ *can be considered as an M-functional with the criterion function* $\rho(x) = |x|$, *and the sample median* $T_n = \tilde{X}_n$ *is a solution of the minimization*

$$\sum_{i=1}^{n} |X_i - \theta| := \min, \quad \theta \in \mathbb{R}$$

To derive the influence function of the median, assume that the probability distribution P has a continuous distribution function F, strictly increasing in interval $(a, b), -\infty \leq a < b \leq \infty$, *and differentiable in a neighborhood of* $\tilde{X}$. *Let* F_t *be a distribution function of the contaminated distribution* $P_t = (1 - t)P + t\delta_x$. *Median* $T(P_t)$ *is a solution of the equation* $F_t(u) = \frac{1}{2}$, *i.e.,*

$$(1 - t)F(T(P_t)) + tI[x < T(P_t)] = \frac{1}{2}$$

that leads to

$$T(P_t) = \begin{cases} F^{-1}\left(\frac{1}{2(1-t)}\right) & \dots \quad x > T(P_t) \\ F^{-1}\left(\frac{1-2t}{2(1-t)}\right) & \dots \quad x \leq T(P_t) \end{cases}$$

The function $T(P_t)$ *is continuous at* $t = 0$, *because* $T(P_t) \to \tilde{X} = T(P)$ *as* $t \to 0$; *using the following expansions around* $t = 0$

$$\frac{1}{2(1-t)} = \frac{1}{2} + \frac{t}{2} + \mathcal{O}(t^2) \quad and \quad \frac{1-2t}{2(1-t)} = \frac{1}{2} - \frac{t}{2} + \mathcal{O}(t^2)$$

we obtain

$$\lim_{t \to 0} \frac{1}{t}[T(P_t) - F^{-1}(\tfrac{1}{2})] = \frac{1}{2} \, \text{sign} \, (x - F^{-1}(\tfrac{1}{2})) \left[\frac{dF^{-1}(u)}{du}\right]_{u=\frac{1}{2}}$$

and this, in turn, leads to the influence function of the median

$$IF(x; \tilde{X}, F) = \frac{\text{sign } (x - \tilde{X})}{2f(\tilde{X})} \tag{3.15}$$

The influence function of the median is bounded, hence the median is robust, while the expected value is non-robust. The breakdown point of the median is $\varepsilon^* = \frac{1}{2}$, and its global sensitivity $\gamma^* = \frac{1}{2f(\tilde{X})}$, ($\gamma^* = 1.253$ for the standard normal distribution $N(0,1)$).

By (3.15), $\mathbb{E}(IF(x; \tilde{X}, P))^2 = \frac{1}{4f^2(\tilde{X})} = \text{const}$ and we can show that the sequence $\sqrt{n}(\tilde{X}_n - \tilde{X})$ is asymptotically normally distributed,

$$\mathcal{L}\{\sqrt{n}(\tilde{X}_n - \tilde{X})\} \to \mathcal{N}\left(0, \frac{1}{4f^2(\tilde{X})}\right)$$

as $n \to \infty$. Especially, if F is the distribution function of the normal distribution $\mathcal{N}(\mu, \sigma^2)$, then $f^2(\tilde{X}) = f^2(\mu) = \frac{1}{2\pi\sigma^2}$ and

$$\mathcal{L}\{\sqrt{n}(\tilde{X}_n - \tilde{X})\} \to \mathcal{N}\left(0, \frac{\pi\sigma^2}{2}\right).$$

In R a function `median` *can be used to compute the sample median.*

(c) Maximal likelihood estimator of parameter θ of the probability distribution with density $f(x, \theta)$:

$$\rho(x, T(P)) = -\log f(x, T(P))$$

$$\psi(x, T(P)) = -\frac{\partial}{\partial\theta} \log f(x, \theta)\Big|_{\theta=T(P)}$$

$$IF(x; T, P) = \frac{1}{\mathcal{I}_f(T(P))} \cdot \frac{\dot{f}(x, T(P))}{f(x, T(P))}$$

where

$$\dot{f}(x, T(P)) = \frac{\partial}{\partial\theta} f(x, \theta)\Big|_{\theta=T(P)}$$

and

$$\mathcal{I}_f(T(P)) = \int_{\mathcal{X}} \left[\frac{\partial}{\partial\theta} \log f(x, \theta)\Big|_{\theta=T(P)}\right]^2 f(x, T(P))dx$$

is the Fisher information of distribution f at the point $\theta = T(P)$.

3.2.2 Choice of function ψ

The M-estimator is determined by the choice of the criterion function ρ or of its derivative ψ. If the location parameter coincides with the center of symmetry of the distribution, we choose ρ symmetric around zero and hence ψ odd.

The influence function of an M-estimator is proportional to $\psi(x - T(P))$ (see (3.10)). Hence, a robust M-estimator should be generated by a bounded ψ. Let us describe the most well-known and the most popular types of functions ψ (and ρ), that we can find in the literature.

The *expected value* is an M-functional generated by a linear, and hence unbounded function ψ. The corresponding M-estimator is the sample mean $\bar{X}_n$, which is the maximal likelihood estimator of the location parameter of the normal distribution. However, this functional is closely connected with the normal distribution and is highly non-robust. If we look for an M-estimator of the location parameter of a distribution not very far from the normal distribution, but possibly containing an ε ratio of nonnormal data, more precisely, that belongs to the family

$$\mathcal{F} = \{F : F = (1 - \varepsilon)\Phi + \varepsilon H\}$$

where H runs over symmetric distribution functions, we should use the function ψ, proposed and motivated by P. J. Huber (1964). This function is linear in a bounded segment $[-k, k]$, and constant outside this segment, see Figure 3.1.

$$\psi_H(x) = \begin{cases} x & \dots \quad |x| \leq k \\ k \, \text{sign} \, x & \dots \quad |x| > k \end{cases} \tag{3.16}$$

where $k > 0$ is a fixed constant, connected with ε through the following identity:

$$2\Phi(k) - 1 + \frac{2\Phi'(k)}{k} = \frac{1}{1 - \varepsilon} \tag{3.17}$$

The corresponding M-estimator is very popular and is often called the *Huber estimator* in the literature. It has a bounded influence function proportional to ψ_H (following from (3.10)), the breakdown point $\varepsilon^* = \frac{1}{2}$, the global sensitivity $\gamma^* = \frac{k}{2F(k)-1}$, and the tail-behavior measure $\lim_{a \to \infty} B(a, T_n, F) = \frac{1}{2}$ both for distributions with exponential and heavy tails. Thus, it is a robust estimator of the center of symmetry, insensitive to the extreme and outlying observations. As Huber proved in 1964, an estimator, generated by the function (3.16), is minimaximally robust for a contaminated normal distribution, while the value k depends on the contamination ratio. An interesting and natural question is whether there exists a distribution F such that the Huber M-estimator is the maximal likelihood estimator of θ for $F(x - \theta)$, i.e., such that ψ is the likelihood function for F. Such a distribution really exists, and its density is normal in interval $[-k, k]$, and exponential outside.

In practice, the studentized versions of M-estimators are usually computed, see Section 3.5. So examples in R of all the M-estimators are postponed to that section.

3.2.3   Other choices of ψ

Some authors recommend reducing the effect of outliers even more and choosing a redescending function $\psi(x)$, tending to 0 as $x \to \pm\infty$, eventually vanishing outside a bounded interval containing 0. Such is the *likelihood function of the Cauchy distribution*, see Figure 3.2.

$$\psi_C(x) = -\frac{f'(x)}{f(x)} = \frac{2x}{1+x^2} \tag{3.18}$$

where $f(x) = \frac{1}{\pi(1+x^2)}$ is the density of the Cauchy distribution.

Another example is the *Tukey biweight function* (also known as bisquare function) (Figure 3.3),

$$\psi_T(x) = \begin{cases} x\left[1 - \left(\frac{x}{k}\right)^2\right]^2 & \dots \quad |x| \le k \\ 0 & \dots \quad |x| > k \end{cases} \tag{3.19}$$

and the *Andrews sinus function* (Figure 3.4),

$$\psi_A(x) = \begin{cases} \sin\frac{x}{k} & \dots \quad |x| \le k\pi \\ 0 & \dots \quad |x| > k\pi \end{cases} \tag{3.20}$$

Hampel (1974) proposed a continuous, piecewise linear function ψ (see Figure 3.5), vanishing outside a bounded interval:

$$\psi_{HA}(x) = \begin{cases} |x|\ \mathrm{sign}\ x & \dots \quad |x| < a \\ a\ \mathrm{sign}\ x & \dots \quad a \le |x| < b \\ \frac{c-|x|}{c-b}\ a\ \mathrm{sign}\ x & \dots \quad b \le |x| < c \\ 0 & \dots \quad |x| > c \end{cases} \tag{3.21}$$

In the robustness literature we can also find the *skipped mean*, generated by the function (Figure 3.6)

$$\psi^*(x) = \begin{cases} x & \dots \quad |x| \le k \\ 0 & \dots \quad |x| > k \end{cases} \tag{3.22}$$

or the *skipped median*, generated by the function (Figure 3.7)

$$\tilde{\psi}(x) = \begin{cases} -1 & \dots \quad -k \le x < 0 \\ 0 & \dots \quad |x| > k \\ 1 & \dots \quad 0 \le x \le k \end{cases} \tag{3.23}$$

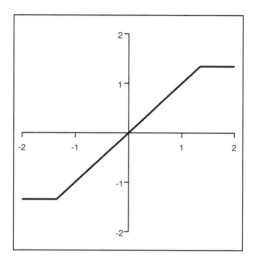

Figure 3.1 *Huber function ψ_H with $k = 1.345$.*

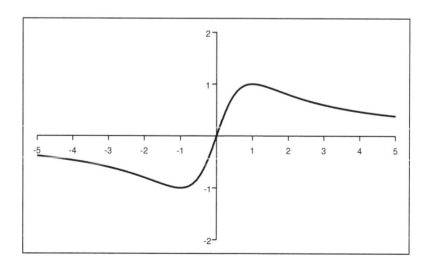

Figure 3.2 *Cauchy function ψ_C.*

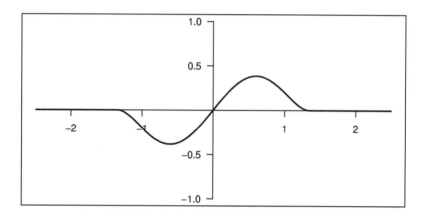

Figure 3.3 *Tukey biweight function ψ_T with $k = 1.345$.*

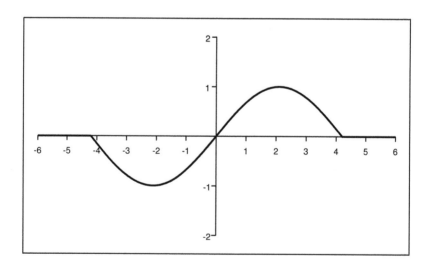

Figure 3.4 *Andrews sinus function ψ_A with $k = 1.339$.*

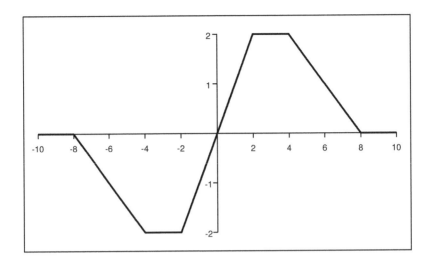

Figure 3.5 *Hampel function ψ_{HA} with $a = 2$, $b = 4$, $c = 8$.*

The redescending functions are not monotone, and their corresponding primitive functions ρ are not convex. Besides the global minimum, the function $\sum_{i=1}^{n} \rho(X_i - \theta)$ can have local extremes, inducing further roots of the equation $\sum_{i=1}^{n} \psi(X_i - \theta) = 0$. Moreover, the functions ψ generating the skipped mean and the skipped median have jump discontinuities, and hence the equation $\sum_{i=1}^{n} \psi(X_i - \theta) = 0$ generally has no solution; the corresponding M-estimator must be calculated as a global minimum of the function $\sum_{i=1}^{n} \rho(X_i - \theta)$.

3.3 Finite sample minimax property of M-estimator

The Huber estimator is asymptotically minimax over the family of contaminated normal distributions. We shall now illustrate another finite sample minimax property of the Huber M-estimator, proved by Huber in 1968.

Consider a random sample from a population with distribution function $F(x - \theta)$ where both F and θ are unknown, and assume that F belongs to the Kolmogorov ε-neighborhood of the standard normal distribution, i.e.,

$$F \in \mathcal{F} = \{F : \sup_{x \in \mathbf{R}} |F(x) - \Phi(x)| \leq \varepsilon\} \qquad (3.24)$$

where Φ is the standard normal distribution function. Fix an $a > 0$ and consider the inaccuracy measure of an estimator T of θ :

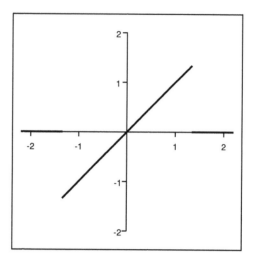

Figure 3.6 *Skipped means function ψ^* with $k = 1.345$.*

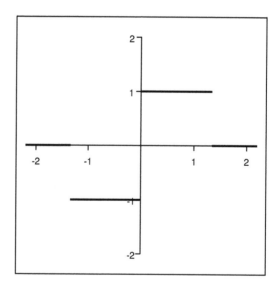

Figure 3.7 *Skipped medians function $\tilde{\psi}^*$ with $k = 1.345$.*

$$\sup_{F \in \mathcal{F}, \theta \in \mathbf{R}} \{ P_\theta | T - \theta) | > a \} \tag{3.25}$$

Let T_H be a slightly modified, randomized Huber estimator:

$$T_H = \begin{cases} T^* & \text{with probability} \quad \frac{1}{2} \\ T^{**} & \text{with probability} \quad \frac{1}{2} \end{cases} \tag{3.26}$$

where

$$T^* = \sup\{t : \sum_{i=1}^{n} \psi_H(X_i - t) \geq 0\}$$

$$\tag{3.27}$$

$$T^{**} = \inf\{t : \sum_{i=1}^{n} \psi_H(X_i - t) \leq 0\}$$

ψ_H is the Huber function (3.16), and the randomization does not depend on $X_1, \ldots, X_n$.

Then T_H is translation equivariant, i.e., satisfies (3.11). We shall show that T_H minimizes the inaccuracy (3.25) among all translation equivariant estimators of θ. To be more precise, let us formulate it as a theorem. The sketch of the proof of Theorem 3.1 can be omitted on the first reading.

Theorem 3.1 *Assume that the bound k in (3.12) is connected with ε and with $a > 0$ in (3.25) through the following identity:*

$$e^{-2ak}[\Phi(a - k) - \varepsilon] + \Phi(a + k) - \varepsilon = 1 \tag{3.28}$$

where Φ is the standard normal distribution function. Then the estimator T_H defined in (3.26) and (3.27) minimizes the inaccuracy (3.25) in the family of translation equivariant estimators of θ.

Sketch of the proof: The main idea is to construct a minimax test of the hypothesis that the parameter equals $-a$, against the alternative that it equals $+a$. The estimator T_H will be an inversion of this minimax test.

Let Φ be the standard normal distribution function and $\phi(x)$, $x \in \mathbb{R}$ be its density. Moreover, denote

$$p_-(x) = \phi(x - a), \quad p_+(x) = \phi(x + a), \quad x \in \mathbb{R}$$

as the shifted normal densities, $\Phi_-(x) = \Phi(x - a)$ and $\Phi_+(x) = \Phi(x + a)$ their distribution functions, and P_- and P_+ the corresponding probability distributions. Then $\Phi_-(x) < \Phi_+(x) \, \forall x$, and the likelihood ratio

$$\frac{p_-(x)}{p_+(x)} = e^{2ax} \tag{3.29}$$

is strictly increasing in x.

Introduce two families of distribution functions:

$$\mathcal{F}_- = \{G \in \mathcal{F} : G(x) \le \Phi(x-a) + \varepsilon \ \ \forall x \in \mathbb{R}\}$$

$$\mathcal{F}_+ = \{G \in \mathcal{F} : G(x) \ge \Phi(x+a) - \varepsilon \ \ \forall x \in \mathbb{R}\}$$

(3.30)

We can assume that $\mathcal{F}_- \cap \mathcal{F}_+ = \emptyset$, which is true for sufficiently small ε. We shall look for the minimax test of the hypothesis $\mathbf{H} : F \in \mathcal{F}_-$ against the alternative $\mathbf{K} : F \in \mathcal{F}_+$. This test will be the likelihood ratio test of two least favorable distributions of families $\mathcal{F}_-$ and $\mathcal{F}_+$, respectively. We shall show that the least favorable distributions have the densities:

$$g_-(x) = \begin{cases} [p_+(x) + p_-(x)](1 + e^{2ak})^{-1} & \dots & x < -k \\ p_-(x) & \dots & |x| \le k \\ [p_+(x) + p_-(x)](1 + e^{-2ak})^{-1} & \dots & x > k \end{cases}$$

(3.31)

and

$$g_+(x) = \begin{cases} [p_+(x) + p_-(x)](1 + e^{-2ak})^{-1} & \dots & x < -k \\ p_+(x) & \dots & |x| \le k \\ [p_+(x) + p_-(x)](1 + e^{2ak})^{-1} & \dots & x > k \end{cases}$$

(3.32)

Denote G_-, G_+ as the distribution functions and Q_-, Q_+ as the probability distributions corresponding to densities g_-, g_+, respectively. We can easily verify that the log-likelihood ratio of g_- and g_+ is connected with ψ_H in the following way:

$$\ln \prod_{i=1}^n \frac{g_-(X_i)}{g_+(X_i)} = 2a \sum_{i=1}^n \psi_H(X_i)$$

and the likelihood ratio test of the hypothesis that the true distribution is g_+ against g_- with minimax risk α rejects the hypothesis for large values of the likelihood ratio, i.e., when $\sum_{i=1}^n \psi_H(X_i) > K$ for a suitable K. Mathematically such a test is characterized by a *test function* $\zeta(\mathbf{x})$ that is the probability that the test rejects the hypothesis under observation $\mathbf{x}$:

$$\zeta(\mathbf{x}) = \begin{cases} 1 & \dots & \sum_{i=1}^n \psi_H(X_i) > K \\ \gamma & \dots & \sum_{i=1}^n \psi_H(X_i) = K \\ 0 & \dots & \sum_{i=1}^n \psi_H(X_i) < K \end{cases}$$

(3.33)

where K is determined so that

$$Q_+(\sum_{i=1}^n \psi_H(X_i) > K) = \alpha, \quad Q_-(\sum_{i=1}^n \psi_H(X_i) > K) = 1 - \alpha$$

(3.34)

and $\alpha \in (0, \frac{1}{2})$ is the minimax risk; from the symmetry we conclude that $K = 0$ and $\gamma = \frac{1}{2}$.

It remains to show that G_-, G_+ is really the least favorable pair of distribution functions for families (3.30), and that the test (3.33) is minimax. But it follows from the inequalities

$$Q'_- \left(\frac{g_-(X)}{g_+(X)} > t \right) \geq Q_- \left(\frac{g_-(X)}{g_+(X)} > t \right)$$

$$\tag{3.35}$$

$$Q'_+ \left(\frac{g_-(X)}{g_+(X)} > t \right) \leq Q_+ \left(\frac{g_-(X)}{g_+(X)} > t \right)$$

that hold for all $Q'_- \in \mathcal{F}_-$ and $Q'_+ \in \mathcal{F}_+$ and $t > 0$. Indeed, it is trivially true for $\frac{1}{2a} \ln t < -k$ and $\frac{1}{2a} \ln t > k$, and for $-k \leq \frac{1}{2a} \ln t \leq k$ it follows from (3.30), (3.31) and (3.32).

If the distribution P of X_i belongs to $\mathcal{F}_-$, $i = 1, \ldots, n$ then it follows from (3.35) that the likelihood ratio $\prod_{i=1}^{n} \frac{g_-(X_i)}{g_+(X_i)}$ is the stochastically smallest provided the X_i are identically distributed with density g_-. Analogously, if the distribution P of X_i belongs to $\mathcal{F}_+$, $i = 1, \ldots, n$ then the likelihood ratio $\prod_{i=1}^{n} \frac{g_-(X_i)}{g_+(X_i)}$ is the stochastically largest provided the X_i are identically distributed with density g_+. Thus the test (3.33) minimizes

$$\max \left[\sup_{G' \in \mathcal{F}_+} \mathbb{E}_{G_+}(\zeta), \ \sup_{G' \in \mathcal{F}_-} \mathbb{E}_{G_-}(1 - \zeta) \right] \tag{3.36}$$

hence it is really minimax.

If the distribution of $X - \theta$ belongs to $\mathcal{F}$, then that of $X - \theta - a$ and that of $X - \theta + a$ belong to $\mathcal{F}_+$ and $\mathcal{F}_-$, respectively, and

$$P_\theta \left(T_H(\mathbf{X}) > \theta + a \right) = P_0 \left(T_H(\mathbf{X}) > a \right)$$

$$= \tfrac{1}{2} P_0 \left(T^*(\mathbf{X}) > a \right) + \tfrac{1}{2} P_0 \left(T^{**}(\mathbf{X}) > a \right)$$

$$= \tfrac{1}{2} P_0 \left(T^*(X_1 - a, \ldots, X_n - a) > 0 \right) + \tfrac{1}{2} P_0 \left(T^{**}(X_1 - a, \ldots, X_n - a) > 0 \right)$$

$$\leq \tfrac{1}{2} P_0 \left\{ \sum_{i=1}^{n} \psi_H(X_i - a) > 0 \right\} + \tfrac{1}{2} P_0 \left\{ \sum_{i=1}^{n} \psi_H(X_i - a) \geq 0 \right\}$$

$$= \mathbb{E}_{P_0} \left(\zeta(X_1 - a, \ldots, X_n - a) \right) \leq \mathbb{E}_{Q_+} \left(\zeta(X_1 - a, \ldots, X_n - a) \right) = \alpha$$

as it follows from (3.34) and (3.36).

Similarly we verify that

$$P_\theta \left(T_H(\mathbf{X}) < \theta - a \right) \leq \alpha$$

Let T now be a translation equivariant estimator. Because the distributions Q_+ and Q_- are absolutely continuous, T has a continuous distribution function both under Q_+ and Q_- (see Problem **3.4**), hence $Q_+(T(\mathbf{X}) = 0) = Q_-(T(\mathbf{X}) = 0) = 0$.

Then T induces a test of $\mathcal{F}_+$ against $\mathcal{F}_-$ rejecting when $T(\mathbf{X}) > 0$, and because the test based on the Huber estimator is minimax with the minimax risk α, we conclude that

$$\sup_{Q'_+ \in \mathcal{F}_+, Q'_- \in \mathcal{F}_-} \max \left[Q'_+(T > 0), Q'_-(T < 0)) \right] \geq \alpha$$

hence no equivariant estimator can be better than T_H.   □

3.4   Moment convergence of M-estimators

Summarizing the conditions imposed on a good estimator, it is desirable to have an M-estimator T_n with a bounded influence function and with a breakdown point $1/2$, which estimates θ consistently with the rate of consistency $\sqrt{n}$ and $\sqrt{n}(T_n - \theta)$ that has an asymptotic normal distribution. The asymptotic distribution naturally has finite moments; however, we also wish T_n to have finite moments tending to the moments of the asymptotic distribution. Otherwise, we would welcome the uniform integrability of the sequence $\sqrt{n}(T_n - \theta)$ and its powers.

Indeed, we can prove the moment convergence of M-estimators for a broad class of bounded ψ-functions and under some conditions on density f. For an illustration, we shall prove that under the following conditions (**A.1**) and (**A.2**). The conditions can be still weakened, but (**A.1**) and (**A.2**) already cover a broad class of M-estimators with a bounded influence. This was first proved by Jurečková and Sen (1982).

(**A.1**) $X_1, \ldots, X_n$ is a random sample from a distribution with density $f(x - \theta)$, where f is positive, symmetric, absolutely continuous and nonincreasing for $x \geq 0$; we assume that f has positive and finite Fisher information,

$$0 < \mathcal{I}(f) = \int_{\mathbb{R}} \left(\frac{f'(x)}{f(x)} \right)^2 dF(x) < \infty$$

and that there exists a positive number ℓ (not necessarily an integer or ≥ 1) such that

$$\mathbb{E}|X_1|^\ell = \int_{\boldsymbol{R}} |x|^\ell dF(x) < \infty$$

(**A.2**) ψ is nondecreasing and skew-symmetric, $\psi(x) = -\psi(-x)$, $x \in \boldsymbol{R}$, and

$$\psi(x) = \psi(c) \cdot \text{sign } x \quad \text{for } |x| > c, \quad c > 0$$

Moreover, ψ can be decomposed into absolutely continuous and step components, i.e., $\psi(x) = \psi_1(x) + \psi_2(x)$, $x \in \mathbb{R}$, where ψ_1 is absolutely continuous inside $(-c, c)$ and ψ_2 is a step function that has a finite number of jumps inside $(-c, c)$, i.e.,

$$\psi_2(x) = b_j \ldots d_{j-1} < x < d_j, \ j = 1, \ldots, m+1$$
$$d_0 = -c, \ d_{m+1} = c$$

Theorem 3.2 *For every* $r > 0$*, there exists* $n_r < \infty$ *such that, under conditions* **(A.1)** *and* **(A.2)**,

$$\mathbb{E}_\theta \left[n^{\frac{r}{2}} |T_n - \theta|^r \right] < \infty, \quad \text{uniformly in } n \geq n_r \tag{3.37}$$

Moreover,

$$\lim_{n \to \infty} \mathbb{E}_\theta \left(\sqrt{n} |T_n - \theta| \right)^r = \nu^r \int_{\mathbf{R}} |x|^r d\Phi(x) \tag{3.38}$$

and, especially,

$$\lim_{n \to \infty} \mathbb{E}_\theta \left(\sqrt{n} |T_n - \theta| \right)^{2r} = \nu^{2r} \frac{(2r)!}{2^r r!} \tag{3.39}$$

for $r = 1, 2, \ldots,$ *where*

$$\nu^2 = \frac{\sigma^2}{\gamma^2}, \qquad \sigma^2 = \int_{\mathbf{R}} \psi^2(x) dF(x)$$

$$\tag{3.40}$$

$$\gamma = \int_{\mathbf{R}} \psi_1'(x) dF(x) + \sum_{j=1}^{m} (b_j - b_{j-1}) f(d_j) \ (> 0)$$

and Φ *is the standard normal distribution function.*
Furthermore, $n^{1/2}(T_n - \theta)$ *is asymptotically normally distributed*

$$\mathcal{L}\{n^{1/2}(T_n - \theta)\} \to \mathcal{N}(0, \nu^2) \tag{3.41}$$

Sketch of the proof. We can put $\theta = 0$, without loss of generality.
First, because F has the finite ℓth absolute moment, then

$$\max_{x \in \mathbb{R}} \{|x| F(x)(1 - F(x))\} = C_\ell < \infty \tag{3.42}$$

and

$$\int_{\mathbb{R}} [F(x)(1 - F(x))]^\lambda dx < \infty \quad \forall \lambda > \frac{1}{\ell} > 0 \tag{3.43}$$

Let $a_1 > c > 0$, where c comes from condition **(A.2)**. Then

$$\mathbb{E}\left\{ n^{\frac{r}{2}} |T_n|^r \right\} = \int_0^\infty r t^{r-1} P(\sqrt{n} |T_n| > t) dt \tag{3.44}$$

$$= \left\{ \int_0^{a_1 \sqrt{n}} + \int_{\sqrt{n}}^\infty \right\} r t^{r-1} P(\sqrt{n} |T_n| > t) dt = I_{n1} + I_{n2}$$

We shall first estimate the probability

$$P(\sqrt{n}|T_n| > t) = 2P(\sqrt{n}T_n > t)$$

$$\le 2P\Big\{\frac{1}{n}\sum_{i=1}^{n}\psi(X_i - tn^{-\frac{1}{2}}) - \mathbb{E}\Big[\frac{1}{n}\sum_{i=1}^{n}\psi(X_i - tn^{-\frac{1}{2}})\Big]$$

$$\ge -\mathbb{E}[\psi(X_1 - tn^{-\frac{1}{2}})]\Big\}$$

where we use the inequalities

$$-\mathbb{E}[\psi(X_1 - tn^{-\frac{1}{2}}) = -\mathbb{E}[\psi(X_1 - tn^{-\frac{1}{2}} - \psi(X_1)] \qquad (3.45)$$

$$= \int_{-c}^{c}[F(x + tn^{-\frac{1}{2}}) - F(x)]d\psi(x)$$

$$= \int_{0}^{c}[F(x + tn^{-\frac{1}{2}}) - F(x - tn^{-\frac{1}{2}})]d\psi(x)$$

$$\ge 2tn^{-\frac{1}{2}}f(c + tn^{-\frac{1}{2}})[\psi(c) - \psi(0)]$$

$$= 2tn^{-\frac{1}{2}}f(c + a_1)\psi(c), \quad \forall t \in (0, a_1\sqrt{n})$$

based on the facts that $f(x) \searrow 0$ as $x \to \infty$ and $\psi(x) + \psi(-x) = 0$, $F(x) + F(-x) = 1$ $\forall x \in \mathbb{R}$. Hence, by (3.44) and (3.45), for $0 < t < a_1\sqrt{n}$,

$$P(\sqrt{n}|T_n| > t) \le 2P\Big\{\frac{1}{n}\sum_{i=1}^{n}Z_{ni} \ge 2tn^{-\frac{1}{2}}f(c + a_1)\psi(c)\Big\}$$

where

$$Z_{ni} = \psi\Big(X_i - tn^{-\frac{1}{2}}\Big) - \mathbb{E}\psi\Big(X_i - tn^{-\frac{1}{2}}\Big), \quad i = 1, \ldots, n$$

are independent random variables with means 0, bounded by $2\psi(c)$. Thus we can use the Hoeffding inequality (Theorem 2 in Hoeffding (1963)) and obtain for $0 < t < a_1\sqrt{n}$

$$P(\sqrt{n}|T_n| > t) \le 2\exp\{-a_2 t^2\} \qquad (3.46)$$

where $a_2 \ge 2f^2(c + a_1)\psi^2(c)$. Hence,

$$I_{n1} \le 2r \int_{0}^{a_1\sqrt{n}} \exp\{-a_2 t^2\}t^{r-1}dt \le 2r \int_{0}^{\infty} \exp\{-a_2 t^2\}t^{r-1}dt < \infty \quad (3.47)$$

On the other hand, if $t \ge a_1\sqrt{n}$, then for $n = 2m \ge 2$,

$$P(\sqrt{n}|T_n| > t) = 2P(\sqrt{n}T_n > t) \le 2P(X_{n:m+1} \ge -c + tn^{-\frac{1}{2}}) \qquad (3.48)$$

$$\le 2n \binom{n-1}{m-1} \int_{F(-c+tn^{-\frac{1}{2}})}^{1} u^m(1-u)^{n-m-1}du \le 2[q(F(-c + tn^{-\frac{1}{2}}))]^n$$

where $X_{n:1} \le \ldots \le X_{n:n}$ are the order statistics and

$$q(u) = 4u(1 - u) \le 1, \quad 0 \le u \le 1$$

Actually, $F(-c + tn^{-\frac{1}{2}}) > \frac{1}{2}$ for $t \geq a_1\sqrt{n}$, and

$$2n \begin{pmatrix} n-1 \\ m-1 \end{pmatrix} \int_A^1 u^m(1-u)^{n-m-1}du \leq 2[q(A)]^n \quad \text{for } A > \frac{1}{2}$$

which can be proved again with the aid of the Hoeffding inequality. If $n = 2m + 1$, we similarly get

$$P(\sqrt{n}|T_n| > t) \leq 2n \begin{pmatrix} n-1 \\ m \end{pmatrix} \int_{F(-c+tn^{-\frac{1}{2}})}^1 u^m(1-u)^{n-m}du$$

$$\leq 2[q(F(-c+tn^{-\frac{1}{2}}))]^n \tag{3.49}$$

Finally, using (3.42)–(3.44), (3.46), (3.48)–(3.49), we obtain

$$I_{n2} \leq 2r \int_{a_1\sqrt{n}} t^{r-1}[q(F(-c+tn^{-\frac{1}{2}}))]^n dt = 2rn^{\frac{r}{2}} \int_{a_1} x^{r-1}[q(F(-c+x))]^n dx$$

$$\leq 2r(C_\ell)^{\frac{r-1}{\ell}} \left\{ n^{\frac{r}{2}}[q(F(-c+a_1))]^{n-[\frac{r}{\ell}]-1-\lambda} \right\} \int_{a_1-c} [4F(y)(1-F(y))]^\lambda dy < \infty$$

for $n \geq [\frac{r}{\ell}] + 1$, and $\lim_{n\to\infty} I_{n2} = 0$. This, combined with (3.47), proves (3.37).

It remains to prove the moment convergence (3.39). Under the conditions **(A.1)–(A.2)**, the *M*-estimator admits the asymptotic representation

$$\gamma\sqrt{n}(T_n - \theta) = n^{-\frac{1}{2}}\sum_{i=1}^n \psi(X_i - \theta) + O_p(n^{-\frac{1}{4}})$$

proved by Jurečková (1980). Since $\mathbb{E}_\theta\psi(X_1 - \theta) = 0$ and ψ is bounded, all moments of $\psi(X_1 - \theta)$ exist. Hence, the von Bahr (1965) theorem on moment convergence of sums of independent random variables applies to $n^{-\frac{1}{2}}\sum_{i=1}^n \psi(X_i - \theta)$, and this further implies (3.40) for any positive integer r, in view of the uniform integrability (3.37). It further extends to any positive real r, because $\left|n^{-\frac{1}{2}}\sum_{i=1}^n \psi(X_i - \theta)\right|^s$ is uniformly integrable for any $s \in [2r-2, 2r]$. The asymptotic normality then follows from the central limit theorem. □

3.5 Studentized *M*-estimators

The *M*-estimator of the shift parameter is translation equivariant but generally it is not scale equivariant (see (3.11)). This shortage can be overcome by using either of the following two methods:

- We estimate the scale simultaneously with the location parameter: e.g., Huber (1981) proposed estimating the scale parameter σ simultaneously with the location parameter θ as a solution of the following system of equations:

$$\sum_{i=1}^{n} \psi_H \left(\frac{X_i - \theta}{\sigma} \right) = 0 \qquad (3.50)$$

$$\sum_{i=1}^{n} \chi \left(\frac{X_i - \theta}{\sigma} \right) = 0 \qquad (3.51)$$

where $\chi(x) = \psi_H^2(x) - \int_{\mathbf{R}} \psi_H^2(y) d\Phi(y)$, ψ_H is the Huber function (3.16), and Φ is the distribution function of the standard normal distribution.

- We can obtain a translation and scale equivariant estimator of θ, if we *studentize* the M-estimator by a convenient scale statistic $S_n(X_1, \ldots, X_n)$ and solve the following minimization:

$$\sum_{i=1}^{n} \rho \left(\frac{X_i - \theta}{S_n} \right) := \min, \quad \theta \in \mathbb{R} \qquad (3.52)$$

However, to guarantee the translation and scale equivariance of the solution of (3.52), our scale statistic should satisfy the following conditions:

(a) $S_n(\boldsymbol{x}) > 0$ *a.e.* for $\boldsymbol{x} \in \mathbb{R}$
(b) $S_n(x_1 + c, \ldots, x_n + c) = S_n(x_1, \ldots, x_n)$, $c \in \mathbb{R}$, $\boldsymbol{x} \in \mathbb{R}^n$
   (translation invariance)
(c) $S_n(cx_1, \ldots, cx_n) = cS_n(x_1, \ldots, x_n)$, $c > 0$, $\boldsymbol{x} \in \mathbb{R}^n$
   (scale equivariance)

Moreover, it is convenient if S_n consistently estimates a statistical functional $S(F)$, so that

$$\sqrt{n}(S_n - S(F)) = \mathcal{O}_p(1) \quad \text{as } n \to \infty \qquad (3.53)$$

Indeed, the estimator defined as in (3.52) is translation and scale equivariant, and the pertaining statistical functional $T(F)$ is defined implicitly as a solution of the minimization

$$\int_{\mathcal{X}} \rho \left(\frac{x - t}{S(F)} \right) dF(x) := \min, \quad t \in \mathbb{R} \qquad (3.54)$$

The functional is Fisher consistent, provided the solution of the minimization (3.54) is unique. If ρ has a continuous derivative ψ, then the estimator equals a root of the equation

$$\sum_{i=1}^{n} \psi \left(\frac{X_i - \theta}{S_n} \right) = 0 \qquad (3.55)$$

If ρ is convex and hence ψ is nondecreasing, but discontinuous at some points or constant on some intervals, we obtain a unique studentized estimator analogously as in (3.12), namely

$$T_n = \frac{1}{2}(T_n^+ + T_n^-)$$

$$T_n^- = \sup\{t : \sum_{i=1}^{n} \psi\left(\frac{X_i - t}{S_n}\right) > 0\} \qquad (3.56)$$

$$T_n^+ = \inf\{t : \sum_{i=1}^{n} \psi\left(\frac{X_i - t}{S_n}\right) < 0\}$$

There is a variety of possible choices of S_n, because there is no universal scale functional. Let us mention some of the most popular choices of the scale statistic S_n:

- *Sample standard deviation:*

$$S_n = \left(\frac{1}{n}\sum_{i=1}^{n}(X_i - \bar{X}_n)^2\right)^{\frac{1}{2}}$$

$$S(F) = (\mathrm{var}_F(X))^{\frac{1}{2}}$$

This functional, being highly non-robust, is used for studentization only in special cases, as in the Student *t*-test under normality. In R the standard function sd computes the sample standard deviation using denominator $n - 1$.

- *Inter-quartile range:*

$$S_n = X_{n:[\frac{3}{4}n]} - X_{n:[\frac{1}{4}n]}$$

where $X_{n:[np]}$, $0 < p < 1$ is the empirical p-quantile of the ordered sample $X_{n:1} \leq \ldots \leq X_{n:n}$. The corresponding functional has the form

$$S(F) = F^{-1}(\tfrac{3}{4}) - F^{-1}(\tfrac{1}{4}).$$

In R we can use a function IQR to compute the inter-quartile range.

- *Median absolute deviation (MAD):*

$$S_n = \mathrm{med}_{1 \leq i \leq n}|X_i - \tilde{X}_n|$$

The corresponding statistical functional $S(F)$ is a solution of the equation

$$F\left(S(F) + F^{-1}(\tfrac{1}{2})\right) - F\left(-S(F) + F^{-1}(\tfrac{1}{2})\right) = \tfrac{1}{2}$$

and $S(F) = F^{-1}(\tfrac{3}{4})$ provided the distribution function F is symmetric around 0 and $F^{-1}(\tfrac{1}{2}) = 0$. Function mad computes the median of the absolute deviations from the center, defaults to the median, and multiplies by a constant. The default value $= 1.4826$ ensures the asymptotically normal consistency.

The influence function of the studentized M-functional in the symmetric model satisfying $F(-x) = 1 - F(x)$, $\rho(-x) = \rho(x)$, with absolutely continuous ψ satisfying $\psi(-x) = -\psi(x)$, $x \in \mathbb{R}$, has the form

$$IF(x, T, F) = \frac{S(F)}{\gamma(F)} \, \psi \left(\frac{x - T(F)}{S(F)} \right)$$

where $\gamma(F) = \int_{\mathbb{R}} \psi' \left(\frac{y}{S(F)} \right) dF(y)$. Hence, the influence function of $T(F)$ in the symmetric model depends on the value of $S(F)$, but not on the influence function of the functional $S(F)$.

In a MASS package there are functions that can compute some of the above mentioned M-estimators. Some others will be computed by our own procedures.

The MASS package also contains many datasets. We choose the dataset chem as an example of illustration of R functions. The dataset contains 24 determinations of copper in wholemeal flour, in parts per million (Analytical Methods Committee (1989), Venables and Ripley (2002)). We start with basic estimators of location (mean, median) and scale (sd, mad, IQR).

```
> library(MASS)
> sort(chem)
 [1]  2.20  2.20  2.40  2.40  2.50  2.70  2.80  2.90  3.03  3.03
[11]  3.10  3.37  3.40  3.40  3.40  3.50  3.60  3.70  3.70  3.70
[21]  3.70  3.77  5.28 28.95
> mean(chem)
[1] 4.280417
> median(chem)
[1] 3.385
> sd(chem)
[1] 5.297396
> mad(chem)
[1] 0.526323
> IQR(chem)
[1] 0.925
>
```

The procedures for location Huber M-estimation are incorporated into the system R. The function huber finds the Huber M-estimator of location with MAD scale (solution of (3.52)) and the function hubers finds the Huber M-estimator for location with scale specified, scale with location specified, or both if neither is specified (solving (3.50) and (3.51)). Both functions are stored in the MASS package.

```
> huber(chem)
$mu
[1] 3.206724
```

```
$s
[1] 0.526323

> hubers(chem)
$mu
[1] 3.205498

$s
[1] 0.673652

> hubers(chem, s = mad(chem)) # the same as huber(chem)
$mu
[1] 3.206724

$s
[1] 0.526323

>
```

We can also use the function `rlm` that fits a robust linear regression model using *M*-estimators and that is also available in the MASS package. By entering only the intercept into the model formula, this function returns the *M*-estimator of location (Huber by default). The psi function can be specified by an argument `psi` with possible values `psi.huber`, `psi.hampel` and `psi.bisquare`. This function by default uses an iterated re-scaled MAD of the residuals from zero as the scale estimate. The least-square fit is used as the initial estimate by default but can be changed by the argument `init`. The algorithm is based on iterated re-weighted least squares.

```
> rlm(chem~1) # k = 1.345 by default
Call:
rlm(formula = chem ~ 1)
Converged in 5 iterations

Coefficients:
(Intercept)
3.205025

Degrees of freedom: 24 total; 23 residual
Scale estimate: 0.734
>
> rlm(chem~1, k = 1.5)
Call:
rlm(formula = chem ~ 1, k = 1.5)
Converged in 5 iterations
```

```
Coefficients:
(Intercept)
3.212203

Degrees of freedom: 24 total; 23 residual
Scale estimate: 0.723
>
> rlm(chem~1,psi=psi.bisquare) # Tukey biweight psi
Call:
rlm(formula = chem ~ 1, psi = psi.bisquare)
Converged in 5 iterations

Coefficients:
(Intercept)
3.167738

Degrees of freedom: 24 total; 23 residual
Scale estimate: 0.741
>
> rlm(chem~1,psi=psi.hampel)
Call:
rlm(formula = chem ~ 1, psi = psi.hampel)
Converged in 3 iterations

Coefficients:
(Intercept)
3.18102

Degrees of freedom: 24 total; 23 residual
Scale estimate: 0.741
>
```

We can see that the solution depends on the tuning constants and on the selected method of scale estimation. In case of a redescending ψ-function, we deal with a non-convex optimization problem, in which case we have multiple local minima, and a good starting point is important, see Example 3.8.

We have also prepared our own procedures for M-estimators of location with Huber (3.16), Cauchy (3.18), Tukey biweight (3.19), Hampel (3.21) and skipped mean ψ-function (3.22). Here is the function hampel for the Hampel M-estimator, the remaining procedures (huber2, cauchy, biweight, skipped.mean) are similar and can be found in Appendix A.

```
hampel <- function (y, a = 2, b = 4, c = 8, initmu = median(y),
                    s = mad(y), iters = FALSE, tol = 1e-08)
{
  require(robustbase)
```

```
y   <- y[!is.na(y)]
n   <- length(y)
mu <- as.numeric(initmu)
s   <- as.numeric(s)
Niter <- 300
Converged <- FALSE
if (s == 0)
  stop("cannot estimate scale: s is zero for this sample")
for (i in 0:Niter) {
if (iters) s <- mad(y,mu) #s ignored, MAD of current residuals
yy   <- Mpsi((y - mu)/s, cc = c(a, b, c), psi="hampel") * s
mu1 <- sum(yy)/n
mu   <- mu + mu1
if (abs(mu1) < tol * s ) {Converged <- TRUE; Niter <- i; break}
                    }
list(mu = mu, s = s, Niter = Niter, Converged = Converged)
}
```

The arguments of the function are the following

    y: vector of data values

    a, b, c: tuning constants

    mu: initial value for the location

    s: the specified scale

    iters: logical value specifying the scale estimation, **FALSE** for fixed scale estimate or **TRUE** for iterated re-scaled MAD of the residuals as in the **rlm** function

    tol: convergence tolerance.

The value of the function is a list of

    mu: the location estimate

    s: the scale estimate

    Niter: number of iterations

    Converged: logical indicating whether the convergence was successful.

The same result can be obtained from the function lmrob in the package robustbase. We also apply the function cauchy.

```
> hampel(chem)
$mu
[1] 3.161176

$s
[1] 0.526323

$Niter
[1] 8
```

```
$Converged
[1] TRUE


> library(robustbase)
> lmrob(chem~1, control = lmrob.control(method="M",psi =
+        "hampel", tuning.psi=c(2,4,8)),init =
+        list(coefficients = median(chem), scale = mad(chem)) )

Call:
lmrob(formula = chem ~ 1, control = lmrob.control(method = "M",
      psi = "hampel",    tuning.psi = c(2, 4, 8)), init =
      list(coefficients = median(chem), scale = mad(chem)))

Coefficients:
(Intercept)
3.161

> cauchy(chem, iters = T)
$mu
[1] 3.216689

$s
[1] 0.7165567

$Niter
[1] 19

$Converged
[1] TRUE
>
```

3.6 S- and τ- estimators, MM-estimators

M-estimators of the location parameter have a high breakdown point, but are
not scale-equivariant. Rousseeuw and Yohai (1984), Yohai (1987), Yohai and
Zamar (1988) and Salibian-Barrera, Willems and Zamar (2008) developed a
class of estimators, related to M-estimators, whose advantages are evident
mainly in the linear model, as we shall see in the next chapter. We shall
briefly illustrate their structure on the location model, where they also have
an interest.

The S-estimator was proposed by Rousseeuw and Yohai (1984). Let
$X_1, \ldots, X_n$ be a random sample from a distribution $F(x - \theta)$, F generally
unknown, symmetric around θ. Denote $r_i(t) = X_i - t$, $i = 1, \ldots, n$ as the
residuals, $t \in \mathbb{R}$. The S-estimator of θ starts from an M-estimator $s(t)$ of
scale generated by a function ρ, which satisfies:

(i) ρ is symmetric, non-decreasing on $(0, m)$ and absolutely continuous with bounded derivative.

(ii) $\rho(0) = 0$, and there exists a $c > 0$ such that ρ is constant on $[c, m)$. For every t, the scale estimate $s(t) = s(r_1(t), \ldots, r_n(t))$ is defined as a solution of

$$\frac{1}{n} \sum_{i=1}^{n} \rho(r_i(t)/s) = K \quad \text{with respect to} \ \ s > 0. \tag{3.57}$$

(iii) with $K = \mathbb{E}_F(\rho(X))$, where distribution function F has a symmetric, unimodal and absolutely continuous density f with f' bounded.

If $\qquad \frac{1}{n} \#\{i : r_i(t) = 0 \ \text{for} \ 1 \leq i \leq n\} < 1 - (K/\rho(c)), \tag{3.58}$

then (3.57) has a unique solution, and this solution is different from 0. If the number of zeros $\geq 1 - (K/\rho(c))$ we define $s(t) = 0$.

(iv) Moreover, if $K > 0$ is fixed, we standardize ρ so that $\frac{K}{\rho(c)} = \frac{1}{2}$.

The *S*-estimator $\widehat{\theta}$ of θ is then defined as

$$\widehat{\theta} = \arg\min\{s(r_1(t), \ldots, r_n(t))\}, t \in \mathbb{R}. \tag{3.59}$$

Hence, the final value of the scale estimator is $\widehat{\sigma} = s(r_1(\widehat{\theta}), \ldots, r_n(\widehat{\theta}))$. Under the above conditions, the *S*-estimator $\widehat{\theta}$ has breakdown point $1/2$:

Lemma 3.1 *An S-estimator generated by function ρ satisfying (i)–(iv) has breakdown point*

$$\varepsilon^* = \frac{1}{n}\left(\left[\frac{n}{2}\right] + 1\right). \tag{3.60}$$

Proof. By (iv), we standardize ρ so that $\rho(c) = 2K$ for fixed K. Without loss of generality, we can assume that ρ is symmetric, non-decreasing on $(0, m)$, $\rho(0) = 0$, $\rho(c) = \frac{1}{2}$ and ρ is constant on $[c, m]$. Otherwise we divide $\rho(\cdot)$ by $2K$.

For a fixed t, denote $|r|_{(1)} \leq \ldots \leq |r|_{(n)}$ as the order statistics corresponding to $|r_1(t)|, \ldots, |r_n(t)|$, and denote $|\tilde{r}(t)|$ as their median. If $|\tilde{r}(t)| \geq \frac{1}{2}$, then (3.57) has no solution and we put $s = \infty$. Let $|\tilde{r}(t)| < \frac{1}{2}$; then there exist $\alpha, \beta > 0$ such that

$$\alpha \cdot |\tilde{r}(t)| \leq s(t) \leq \beta \cdot |\tilde{r}(t)|. \tag{3.61}$$

Indeed, $|r|_{(i)} = \frac{1}{2}$ for $i > k_n$ and $|r|_{(i)} < \frac{1}{2}$ for $i = 1, 2, \ldots, k_n$, where $k_n > \frac{n}{2}$. Moreover, $\rho\left(\frac{|\tilde{r}(t)|}{s(t)}\right) \geq \varepsilon$ for some $\varepsilon > 0$. Then (3.61) is true for $\alpha = \frac{1}{c}$ and $\beta = \frac{1}{\rho^{-1}(\varepsilon)}$; then $\widehat{\theta}$ minimizes the median of $|X_1 - t|, \ldots, |X_n - t|$ with respect to t and has breakdown point (3.60). In fact, $\widehat{\theta}$ is an *L*-estimator, which will be considered in the subsequent sections. $\qquad\square$

Yohai (1987) and Yohai and Zamar (1988) refined the S-estimators, planning to obtain estimators with asymptotic efficiency arbitrarily close to 1 when the errors are independent and identically distributed with a normal distribution, besides their high breakdown points. They constructed MM-estimators and τ estimators, based on improved estimators of scale, starting with an initial location estimator. The final location estimators are based on an improved criterion.

The function rlm from the MASS package can be used to compute this type of estimator. Setting method = "MM" the function uses an initial S-estimator followed by the final M-estimator of location with the Tukey biweight function.

```
> rlm(chem~1, method = "MM")
Call:
rlm(formula = chem ~ 1, method = "MM")
Converged in 5 iterations

Coefficients:
(Intercept)
3.158991

Degrees of freedom: 24 total; 23 residual
Scale estimate: 0.658
>
```

Alternatively, the function lmrob from the robustbase package can be used.

```
> lmrob(chem~1, control = lmrob.control(psi = "hampel"))

Call:
lmrob(formula = chem ~ 1, control =
      lmrob.control(psi = "hampel"))
\--> method = "MM"
Coefficients:
(Intercept)
3.159
>
```

The S-estimator can be extracted the following way:

```
> (MM <- lmrob(chem~1)) #using psi = "bisquare" by default


Call:
lmrob(formula = chem ~ 1)
\--> method = "MM"
Coefficients:
```

```
(Intercept)
3.159

> MM$init.S$coefficients
(Intercept)
3.391392
>
```

Or it can be obtain directly from a function `lmrob.S`:

```
> lmrob.S(x = rep(1,length(chem)), chem, control =
+            lmrob.control(psi = "bisquare"))$coefficients
[1] 3.391392
>
```

3.7 *L*-estimators

L-estimators are based on the ordered observations (order statistics) $X_{n:1} \leq \ldots \leq X_{n:n}$ of the random sample $X_1, \ldots, X_n$. The general *L*-estimator can be written in the form

$$T_n = \sum_{i=1}^{n} c_{ni} h(X_{n:i}) + \sum_{j=1}^{k} a_j h^*(X_{n:[np_j]+1}) \tag{3.62}$$

where $c_{n1}, \ldots, c_{nn}$ and $a_1, \ldots, a_k$ are given coefficients, $0 < p_1 < \ldots < p_k < 1$ and $h(\cdot)$ and $h^*(\cdot)$ are given functions. The coefficients c_{ni}, $1 \leq i \leq n$ are generated by a bounded weight function $J : [0,1] \mapsto \mathbb{R}$ in the following way: either

$$c_{ni} = \int_{\frac{i-1}{n}}^{\frac{i}{n}} J(s)ds, \quad i = 1, \ldots, n \tag{3.63}$$

or approximately

$$c_{ni} = \frac{1}{n} J\left(\frac{i}{n+1}\right), \quad i = 1, \ldots, n \tag{3.64}$$

The first component of the *L*-estimator (3.62) generally involves all order statistics, while the second component is a linear combination of several (finitely many) sample quantiles, see function `quantile` in R. Many *L*-estimators have just the form either of the first or of the second component in (3.62); we speak about *L*-estimators of type I or II, respectively.

The simplest examples of *L*-estimators of location are the sample median $\tilde{X}_n$ and the *midrange*

$$T_n = \frac{1}{2}(X_{n:1} + X_{n:n}).$$

For the computation we can use the standard function `range`.

```
> range(chem)
[1]   2.20 28.95
> mean(range(chem))
[1] 15.575
>
```

The popular *L*-estimators of scale are the *sample range*

$$R_n = X_{n:n} - X_{n:1}$$

```
> diff(range(chem))
[1] 26.75
>
```

and the *Gini mean difference*

$$G_n = \frac{1}{n(n-1)} \sum_{i,j=1}^{n} |X_i - X_j| = \frac{2}{n(n-1)} \sum_{i=1}^{n} (2i - n - 1) X_{n:i}$$

that is implemented, e.g., in package lmomco.

```
> library(lmomco)
> gini.mean.diff(chem)$gini
[1] 2.830906
>
```

The *L*-estimators of type I are more important for applications. Let us consider some of their main characteristics. Let *L*-estimator T_n have an integrable weight function J such that $\int_0^1 J(u)du = 1$. Its corresponding statistical functional is based on the empirical quantile function

$$Q_n(t) = F_n^{-1}(t) = \inf\{x : F_n(x) \geq t\}, \ 0 < t < 1$$

that is the empirical counterpart of the quantile function $Q(t) = F^{-1}(t) = \inf\{x : F(x) \geq t\}, \ 0 < t < 1$ and is equal to

$$Q_n(t) = \begin{cases} X_{n:i} & \dots & \frac{i-1}{n} < t \leq \frac{i}{n}, & i = 1, \dots, n-1 \\ X_{n:n} & \dots & \frac{n-1}{n} < t \leq 1 \end{cases} \tag{3.65}$$

Using the empirical quantile function, we can express the *L*-estimator in an alternative way

$$T_n = \int_0^1 J(s)h\left(Q_n(s)\right) ds \tag{3.66}$$

The corresponding functional has the form

$$T(F) = \int_0^1 J(s)h\left(Q(s)\right) ds \tag{3.67}$$

3.7.1 Influence function of L-estimator

Assume that F is increasing and absolutely continuous with derivative f, and that the function h in (3.62) is absolutely continuous. Denote

$$F_t(y) = (1-t)F(y) + t\delta_x = \begin{cases} (1-t)F(y) & y < x \\ (1-t)F(y) + t & y \geq x \end{cases}$$

as the contamination of F by the distribution function of the constant x, i.e., by

$$\delta_x(y) = \begin{cases} 0 & \text{if } y < x \\ 1 & \text{if } y \geq x \end{cases}$$

Then

$$F_t^{-1}(u) = \begin{cases} F^{-1}\left(\frac{u}{1-t}\right) & u \leq (1-t)F(x) \\ x & (1-t)F(x) < u \leq (1-t)F(x) + t \\ F^{-1}\left(\frac{u-t}{1-t}\right) & u > (1-t)F(x) + t \end{cases}$$

and hence

$$\frac{dF_t^{-1}(u)}{dt} = \begin{cases} \frac{u}{(1-t)^2} \cdot \frac{1}{f\left(F^{-1}\left(\frac{u}{1-t}\right)\right)} & u < (1-t)F(x) \\ \frac{u-1}{(1-t)^2} \cdot \frac{1}{f\left(F^{-1}\left(\frac{u-t}{1-t}\right)\right)} & u > (1-t)F(x) + t \end{cases}$$

This implies that

$$\frac{dT(F_t)}{dt} = \int_0^1 J(u) h'\left(F_t^{-1}(u)\right) \cdot \frac{dF_t^{-1}(u)}{dt} du$$

$$= \int_0^{F_t(x)} \frac{u}{(1-t)^2} \cdot \frac{h'\left(F^{-1}\left(\frac{u}{1-t}\right)\right)}{f\left(F^{-1}\left(\frac{u}{1-t}\right)\right)} J(u) du$$

$$+ \int_{F_t(x)}^1 \frac{u-1}{(1-t)^2} \cdot \frac{h'\left(F^{-1}\left(\frac{u-t}{1-t}\right)\right)}{f\left(F^{-1}\left(\frac{u-t}{1-t}\right)\right)} J(u) du$$

and $t \to 0_+$ leads to the influence function of the functional (3.67):

$$\left. \frac{dT(F_t)}{dt} \right|_{t=0} =$$

$$= \int_0^{F(x)} s \cdot \frac{h'(F^{-1}(u))}{f(F^{-1}(u))} J(u) du + \int_{F(x)}^1 (u-1) \cdot \frac{h'(F^{-1}(u))}{f(F^{-1}(u))} J(u) du$$

$$= \int_0^1 u \cdot \frac{h'(F^{-1}(u))}{f(F^{-1}(u))} J(u) du - \int_{F(x)}^1 \frac{h'(F^{-1}(u))}{f(F^{-1}(u))} J(u) du$$

Hence, the influence function of $T(F)$ is equal to

$$IF(x,T,F) = \int_{-\infty}^{\infty} F(y)h'(y)J(F(y))dy - \int_{x}^{\infty} h'(y)J(F(y))dy \qquad (3.68)$$

Notice that it satisfies the identity

$$\frac{d}{dx}IF(x,T,F) = h'(x)J(F(x))$$

If $h(x) \equiv x$, $x \in \mathbb{R}$, $J(u) = J(1-u)$, $0 < u < 1$ and F is symmetric around 0, the influence function simplifies:

$$
\begin{aligned}
IF(x,T,F) &= \int_{-\infty}^{\infty} F(y)J(F(y))dy - \int_{x}^{\infty} J(F(y))dy \\
&= \int_{0}^{\infty} F(y)J(F(y))dy + \int_{-\infty}^{0} (1-F(-y))J(1-F(-y))dy \\
&\quad - \int_{x}^{\infty} J(F(y))dy \\
&= \int_{0}^{\infty} F(y)J(F(y))dy + \int_{0}^{\infty} (1-F(y))J(F(y))dy \\
&\quad - \int_{x}^{\infty} J(F(y))dy \\
&= \int_{0}^{\infty} J(F(y))dy - \int_{x}^{\infty} J(F(y))dy
\end{aligned}
$$

and hence

$$IF(x,T,F) = \int_{0}^{x} J(F(y))dF(y) \quad \dots \ x \geq 0$$

$$ \tag{3.69}$$

$$IF(-x,T,F) = -IF(x,T,F) \quad \dots \ x \in \mathbb{R}$$

Remark 3.1 *If M-estimator M_n of the center of symmetry is generated by an absolutely continuous function ψ, and L_n is the L-estimator with the weight function $J(u) = c\,\psi'(F^{-1}(u))$, then the influence functions of M_n and L_n coincide.*

3.7.2 Breakdown point of L-estimator

If the L-estimator T_n is *trimmed* in the sense that its weight function satisfies $J(u) = 0$ for $0 < u \leq \alpha$ and $1 - \alpha \leq u < 1$, and $\varepsilon_n^* = \frac{m_n}{n}$ is its breakdown point, then $\lim_{n \to \infty} \varepsilon_n^* = \alpha$.

Example 3.2 *α-trimmed mean* $(0 < \alpha < \frac{1}{2})$ *is the average of the central quantiles:*

$$\bar{X}_{n\alpha} = \frac{1}{n - 2[n\alpha]} \sum_{i=[n\alpha]+1}^{n-[n\alpha]} X_{n:i}$$

hence

$$c_{ni} = \begin{cases} \frac{1}{n-[n\alpha]} & \dots \quad [n\alpha]+1 \le i \le n - [n\alpha] \\ 0 & \dots \quad \text{otherwise} \end{cases}$$

$$J(u) = \frac{1}{1-2\alpha} I[\alpha \le u \le 1 - \alpha]$$

$$T_n = T(F_n) = \frac{1}{1-2\alpha} \int_\alpha^{1-\alpha} F_n^{-1}(u)du$$

$$T(F) = \frac{1}{1-2\alpha} \int_\alpha^{1-\alpha} F^{-1}(u)du$$

The influence function of $\bar{X}_{n\alpha}$ *follows from (3.68):*

$$IF(x,T,F) = \int_{\mathbf{R}} F(y)J(F(y))dy - \int_x^\infty J(F(y))dy$$

$$= \frac{1}{1-2\alpha} \left\{ \int_{F^{-1}(\alpha)}^{F^{-1}(1-\alpha)} F(y)dy - \int_x^\infty I[\alpha < F(y) < 1-\alpha]dy \right\}$$

hence

$$IF(x,T,F) + \mu_\alpha =$$

$$= -\frac{1}{1-2\alpha} \left[\alpha F^{-1}(1-\alpha) - (1-\alpha)F^{-1}(\alpha)\right] I\left[x < F^{-1}(\alpha)\right]$$

$$+ \frac{1}{1-2\alpha} \left[x - \alpha F^{-1}(\alpha) - \alpha F^{-1}(1-\alpha)\right] I\left[F^{-1}(\alpha) \le x \le F^{-1}(1-\alpha)\right]$$

$$+ \frac{1}{1-2\alpha} \left[-\alpha F^{-1}(\alpha) + (1-\alpha)F^{-1}(1-\alpha)\right] I\left[x > F^{-1}(1-\alpha)\right]$$

where

$$\mu_\alpha = \frac{1}{1-2\alpha} \int_\alpha^{1-\alpha} F^{-1}(u)du = \frac{1}{1-2\alpha} \int_{F^{-1}(\alpha)}^{F^{-1}(1-\alpha)} ydF(y)$$

If F is symmetric, then $F^{-1}(u) = -F^{-1}(1-u)$, $0 < u < 1$ and $\mu_\alpha = 0$; then

$$
IF(x, T, F) =
\begin{cases}
-\dfrac{F^{-1}(1-\alpha)}{1-2\alpha} & \cdots & x < -F^{-1}(1-\alpha) \\[3mm]
\dfrac{x}{1-2\alpha} & \cdots & -F^{-1}(1-\alpha) \le x \le F^{-1}(1-\alpha) \\[3mm]
\dfrac{F^{-1}(1-\alpha)}{1-2\alpha} & \cdots & x > F^{-1}(1-\alpha)
\end{cases}
$$

The global sensitivity of the trimmed mean is

$$
\gamma^* = \frac{F^{-1}(1-\alpha)}{1-2\alpha}
$$

Function `mean` *has the argument* `trim`, *the fraction of observations to be trimmed from each end before the computation of mean. So we can use* `mean(x, trim=alpha)` *as the α%-trimmed mean.*

```
> mean(chem, trim = 0.1)
[1] 3.205
>
```

Remark 3.2

If M_n is the Huber estimator of the center of symmetry θ of $F(x-\theta)$, generated by the Huber function ψ_H with $k = F^{-1}(1-\alpha)$ (see (3.16)), then the influence functions of M_n and $\bar{X}_{n\alpha}$ coincide.

Remark 3.3

(i) Let $\varepsilon_n^* = \frac{m_n}{n}$ be the breakdown point of the α-trimmed mean $\bar{X}_{n\alpha}$. Then $\lim_{n\to\infty} \varepsilon_n^* = \alpha$.

(ii) Let $\alpha_n = [k/n]$, $n \ge 3$ and let $B(\bar{X}_{n,\alpha_n}; a)$ be the tail-behavior measure of $\bar{X}_{n,\alpha_n}$, defined in (2.14). Then, if F has exponential tails (2.17),

$$
n - 2k \le \varliminf_{and\to\infty} B(\bar{X}_{n,\alpha_n}; a) \le \varlimsup_{a\to\infty} B(\bar{X}_{n,\alpha_n}; a) \le n - k \qquad (3.70)
$$

and if F has heavy tails (2.19) and $k < \frac{n-1}{2}$, then

$$
\lim_{a\to\infty} B(\bar{X}_{n\alpha}; a) = k + 1 \qquad (3.71)
$$

Example 3.3 *The α-Winsorized mean is an example of an L-estimator of the general form (3.62), that has two components:*

$$\overline{W}_{n\alpha} = T(F_n) \tag{3.72}$$

$$= \frac{1}{n}\left\{[n\alpha]X_{n:[n\alpha]+1} + \sum_{i=[n\alpha]+1}^{n-[n\alpha]} X_{n:i} + [n\alpha]X_{n:n-[n\alpha]}\right\}$$

$$= \alpha F_n^{-1}(\alpha) + \int_{\alpha}^{1-\alpha} F_n^{-1}(u)du + \alpha F_n^{-1}(1-\alpha)$$

$$= \sum_{i=1}^{n} c_{ni}X_{n:i} + \frac{[n\alpha]+1}{n}X_{n:[n\alpha]+1} + \frac{[n\alpha]+1}{n}X_{n:n-[n\alpha]}$$

where

$$c_{ni} = \begin{cases} \frac{1}{n} & \dots & 1+[n\alpha] < i < n-[n\alpha] \\ 0 & \dots & otherwise \end{cases}$$

The extreme quantiles are not trimmed but replaced with quantiles $X_{n:[n\alpha]+1}$ and $X_{n:n-[n\alpha]}$, respectively. For the sake of simplicity, let us consider the model with symmetric distribution function F. The statistical functional $T(F)$ corresponding to $\overline{W}_{n\alpha}$ is

$$T(F) = T_1(F) + T_2(F)$$

$$= \int_{\alpha}^{1-\alpha} F^{-1}(u)du + \alpha F^{-1}(\alpha) + \alpha F^{-1}(1-\alpha)$$

The influence function of $T_1(F)$ follows from (3.68), while the influence function of $T_2(F)$ is a modification of the influence function of the median (3.15) that is the α-quantile with $\alpha = \frac{1}{2}$; thus

$$IF(x, \overline{W}_{n\alpha}, F) =$$

$$= F^{-1}(\alpha) - \frac{\alpha}{f(F^{-1}(\alpha))} \, I[x < F^{-1}(\alpha)]$$

$$+ x \, I[F^{-1}(\alpha) \le x \le F^{-1}(1-\alpha)]$$

$$+ F^{-1}(1-\alpha) + \frac{\alpha}{f(F^{-1}(1-\alpha))} I[x > F^{-1}(1-\alpha)]$$

The global sensitivity of the Winsorized mean is

$$\gamma^* = F^{-1}(\alpha) + \frac{\alpha}{f(F^{-1}(1-\alpha))}$$

*and the limiting breakdown point of $\overline{W}_{n\alpha}$ is $\varepsilon^* = \alpha$. The influence function of
the Winsorized mean has jump points at $F^{-1}(\alpha)$ and $F^{-1}(1-\alpha)$, while the
influence function of the α-trimmed mean is continuous.*

*To compute the α-Winsorized mean we can use the function **winsorized.
mean** from Appendix A, setting the argument **trim** equal to α.*

```
> winsorized.mean(chem, trim = 0.1)
[1] 3.185
>
```

Example 3.4

(i) Sen's weighted mean (Sen (1964)):

$$T_{n,k} = \binom{n}{2k+1}^{-1} \sum_{i=1}^{n} \binom{i-1}{k} \binom{n-i}{k} X_{n:i}$$

*where $0 < k < \frac{n-1}{2}$. Notice that $T_{n,0} = \bar{X}_n$ and $T_{n,k}$ is the sample median if
either n is even and $k = \frac{n}{2} - 1$ or n is odd and $k = \frac{n-1}{2}$.*

*We can find the function **sen.mean** in the package **lmomco**.*

```
> library(lmomco)
> sen.mean(chem, k = trunc(length(chem)*0.10))
$sen
[1] 3.251903

$source
[1] "sen.mean"

>
```

*(ii) The Harrell-Davis estimator of the p-quantile (Harrell and Davis
(1982)):*

$$T_n = \sum_{i=1}^{n} c_{ni} X_{n:i}$$

$$c_{ni} = \frac{\Gamma(n+1)}{\Gamma(k)\Gamma(n-k+1)} \int_{(i-1)/n}^{i/n} u^{k-1}(1-u)^{n-k} du$$

$i = 1, \ldots, n$, where $k = [np]$, $0 < p < 1$.

*(iii) BLUE (asymptotically best linear unbiased estimator) of the location pa-
rameter (more properties described by Blom (1956) and Jung (1955, 1962)).
Let $X_1, X_2, \ldots$ be independent observations with the distribution function
$F(x - \theta)$, where F has an absolutely continuous density f with derivative f'.*

Then the BLUE is the L-estimator with the weight function

$$T_n = \sum_{i=1}^{n} c_{ni} X_{n:i}, \quad c_{ni} = \frac{1}{n} J\left(\frac{i}{n+1}\right), \quad i = 1, \ldots, n$$

$$J(F(x)) = \psi'_f(x), \quad \psi_f(x) = -\frac{f'(x)}{f(x)}, \quad x \in \mathbb{R}$$

3.8   Moment convergence of *L*-estimators

Similar to the case of *M*-estimators, we can prove the moment convergence of *L*-estimators for a broad class of bounded *J*-functions and under some conditions on density *f*. Consider the *L*-estimator

$$T_n = \sum_{i=1}^{n} c_{ni} X_{n:i} \tag{3.73}$$

where $X_{n:1} \leq X_{n:2} \leq \ldots \leq X_{n:n}$ are the order statistics corresponding to observations $X_1, \ldots, X_n$ and

$$c_{ni} = c_{n,n-i+1} \geq 0, \quad i = 1, \ldots, n, \quad \text{and} \quad \sum_{i=1}^{n} c_{ni} = 1$$

and $\tag{3.74}$

$$c_{ni} = c_{n,n-i+1} = 0 \text{ for } i \leq k_n, \text{ where } \lim_{n \to \infty} \frac{k_n}{n} = \alpha_0$$

for some α_0, $0 < \alpha_0 < \frac{1}{2}$. Assume that the independent observations $X_1, \ldots, X_n$ are identically distributed with a distribution function $F(x - \theta)$ such that F has a symmetric density $f(x)$, $f(-x) = f(x)$, $x \in \mathbb{R}$, and f is monotonically decreasing in x for $x \geq 0$, and

$$\sup_{F^{-1}(\alpha_0) \leq x \leq F^{-1}(1-\alpha_0)} |f'(x)| < \infty \tag{3.75}$$

Denote

$$J_n(t) = n c_{ni} \quad \text{for} \quad \frac{i-1}{n} < t \leq \frac{i}{n}, \ i = 1, \ldots, n \tag{3.76}$$

and assume that

$$J_n(t) \to J(t) \text{ a.s. } \forall t \in (0, 1) \tag{3.77}$$

where $J : [0, 1] \mapsto [0, \infty)$ is a symmetric and integrable function, $J(t) = J(1 - t) \geq 0$, $0 \leq t \leq 1$ and $\int_0^1 J(t)dt = 1$. Then (see Huber (1981)), the sequence $\sqrt{n}(T_n - \theta)$ has an asymptotic normal distribution $\mathcal{N}(0, \sigma_L^2)$ where

$$\sigma_L^2 = \int_{\mathbb{R}} \int_{\mathbb{R}} [F(x \wedge y) - F(x)F(y)] J(F(x)) J(F(y)) dx dy \tag{3.78}$$

Theorem 3.3 *Under the conditions (3.74)–(3.78), for any positive integer r,*

$$\lim_{n\to\infty} \mathrm{E}_\theta[\sqrt{n}(T_n - \theta)]^{2r} = \sigma_L^{2r}\frac{(2r)!}{2^r r!} \tag{3.79}$$

We shall only sketch the basic steps of the proof; the detailed proof can be found in Jurečková and Sen (1982). We shall use the following lemma, that follows from the results of Csörgő and Révész (1978):

Lemma 3.2 *Under the conditions of Theorem 3.3, for any $n \geq n_0$, there exists a sequence of random variables $\{Y_{ni}\}_{i=1}^{n+1}$, independent and normally distributed $\mathcal{N}(0,1)$, such that*

$$\left| \sqrt{n}(T_n - \theta) - \frac{1}{\sqrt{n+1}}\sum_{j=1}^{n+1} a_{nj}Y_{nj} \right| = O(n^{-\frac{1}{2}}\log n) \text{ a.s.} \tag{3.80}$$

as $n \to \infty$, where

$$a_{nj} = \sum_{i=j}^{n} b_{ni} - \frac{1}{n+1}\sum_{i=1}^{n+1} b_{ni}, \quad b_{ni} = \frac{c_{ni}}{f\left(F^{-1}\left(\frac{i}{n+1}\right)\right)}, \quad i = 1,\dots,n$$

Proof of Lemma 3.2: Put $\theta = 0$ without a loss of generality. Using Theorem 6 of Csörgő and Révész (1978), we conclude that there exists a sequence of Brownian bridges $\{\mathcal{B}_n(t) : 0 \leq t \leq 1\}$ such that

$$\max_{k_n+1\leq i\leq n-k_n}\left| \sqrt{n}\left[X_{n:i} - F^{-1}\left(\frac{i}{n+1}\right)\right] - \mathcal{B}_n\left(\frac{i}{n+1}\right)f\left(F^{-1}\left(\frac{i}{n+1}\right)\right) \right|$$

$$= O\left(n^{-\frac{1}{2}}\log n\right) \text{ a.s. as } n \to \infty$$

hence

$$\left| \sqrt{n}T_n - \sum_{i=1}^{n} b_{ni}\mathcal{B}_{ni}\left(\frac{i}{n+1}\right) \right| = O\left(n^{-\frac{1}{2}}\log n\right)$$

The process $\left\{(t+1)\mathcal{B}_n\left(\frac{t}{t+1}\right) : t \geq 0\right\}$ is a standard Wiener process W_n on $[0,\infty)$, thus $W_n(k) = \sum_{i=1}^{k} Y_{ni}$, $k = 1, 2, \dots$, where the Y_{ni} are independent random variables with $\mathcal{N}(0,1)$ distributions. $\square$

Sketch of the proof of Theorem 3.3. Because $\sum_{i=1}^{n} c_{ni}F^{-1}\left(\frac{i}{n+1}\right) = 0$, we get by the Jensen inequality (put $\theta = 0$)

$$(\sqrt{n}|T_n|)^{2r} \leq \left(\sum_{i=1}^{n} \sqrt{n}\left|X_{ni} - F^{-1}\left(\frac{i}{n+1}\right)\right| \right)^{2r}$$

$$\leq \sum_{i=k_n+1}^{n-k_n} c_{ni}\left(\sqrt{n}\left|X_{ni} - F^{-1}\left(\frac{i}{n+1}\right)\right| \right)^{2r}$$

hence

$$E_0(\sqrt{n}|T_n|)^{2r} \leq \sum_{i=k_n+1}^{n-k_n} c_{ni} E\left(\sqrt{n}\left|X_{ni} - F^{-1}\left(\frac{i}{n+1}\right)\right|^{2r}\right) < \infty$$

This, together with Lemma 3.2, implies Theorem 3.3.   □

3.9 Sequential *M*- and *L*-estimators, minimizing observation costs

Let T_n be a fixed estimator (e.g., *M*- or *L*-estimator) of θ based on n independent observations $X_1, \ldots, X_n$, and assume that θ is the center of symmetry of distribution function $F(x-\theta)$. Assume that the loss incurred when estimating θ by T_n also includes the expenses; more precisely, let $c > 0$ be the price of one observation and let the global loss be

$$L(T_n, \theta, c) = a(T_n - \theta)^2 + cn \tag{3.81}$$

where $a > 0$ is a constant. The corresponding risk is

$$R_n(T_n, \theta, c) = E_\theta(T_n - \theta)^2 + cn \tag{3.82}$$

Our goal is to find the sample size n minimizing the risk (3.82).

Let us first consider the situation that F is known and that $\sigma_n^2 = nE_\theta(T_n - \theta)^2$ exists for $n \geq n_0$, and that

$$\lim_{n\to\infty} \sigma_n^2 = \sigma^2(F), \quad 0 < \sigma^2(F) < \infty \tag{3.83}$$

Hence, we want to minimize $\frac{1}{n}a\sigma_n + cn$ with respect to n, and if we use the approximation (3.83), the approximate solution $n_0(c)$ has the form

$$n_0(c) \approx \sigma(F)\sqrt{\frac{a}{c}} \tag{3.84}$$

and for the minimum risk we obtain

$$R_n(T_{n_0(c)}, \theta, c) \approx 2\sigma(F)\sqrt{ac} \tag{3.85}$$

where $p(c) \approx q(c)$ means that $\lim_{c\downarrow 0} \frac{q(c)}{p(c)} = 1$. Then obviously $n_0(c) \uparrow \infty$ as $c \downarrow 0$.

If distribution function F is unknown, we cannot know $\sigma^2(F)$ either. But we can still solve the problem sequentially, if there is a sequence $\hat{\sigma}_n$ of estimators of $\sigma(F)$. We set the random sample size *(stopping rule)* N_c, defined as

$$N_c = \min\left\{n \geq n' : n \geq \sqrt{\frac{a}{c}}\left(\hat{\sigma}_n + n^{-\nu}\right), \quad c > 0\right\} \tag{3.86}$$

where n' is an initial sample size and $\nu > 0$ is a chosen number. Then $N_c \uparrow \infty$ with probability 1 as $c \downarrow 0$. The resulting estimator of θ is T_{N_c}, based on N_c

observations $X_1, \ldots, X_{N_c}$. The corresponding risk is

$$R^*(T_n, \theta, c) = a\mathrm{E}\,(T_{N_c} - \theta)^2 + c\mathrm{E}N_c$$

We shall show that, if T_n is either a suitable M-estimator or an L-estimator, then

$$\frac{R^*(T_n, \theta, c)}{R_n(T_{n(c)}, \theta, c)} \to 1 \quad \text{as } c \downarrow 0 \tag{3.87}$$

An interpretation of the convergence (3.87) is that the sequential estimator T_{N_c} is asymptotically (in the sense that $c \downarrow 0$) equally risk-efficient because the optimal non-sequential estimator $T_{n(c)}$ corresponding to the case that $\sigma^2(F)$ is known. Such a problem was first considered by Ghosh and Mukhopadhyay (1979) and later by Chow and Yu (1981) under weaker conditions; they proved (3.87) for T_n as the sample mean and $\hat{\sigma}_n^2$ as the sample variance. Sen (1980) solved the problem for a class of R-estimators (rank-based estimators) of θ.

Let T_n be the M-estimator of θ generated by a nondecreasing and skew-symmetric function ψ, and assume that ψ and F satisfy condition (**A.1**) and (**A.2**) of Section 3.4, put

$$S_n(t) = \sum_{i=1}^{n} \psi(X_i - t)$$

Then T_n is defined by (3.12) and $\sqrt{n}(T_n - \theta)$ is asymptotically normally distributed $\mathcal{N}(0, \sigma^2(\psi, F))$, where

$$\sigma^2(\psi, F) = \frac{\int_{\mathbb{R}} \psi^2(x)dF(x)}{\left(\int_{\mathbb{R}} f(x)d\psi(x)\right)^2}$$

(see Huber (1981)), put

$$s_n^2 = \frac{1}{n}\sum_{i=1}^{n} \psi^2(X_i - T_n)$$

Choose $\alpha \in (0, 1)$ and put

$$M_n^- = \sup\left\{t : n^{-\frac{1}{2}}S_n(t) > s_n^2\Phi^{-1}\left(1 - \frac{\alpha}{2}\right)\right\}$$

$$M_n^+ = \sup\left\{t : n^{-\frac{1}{2}}S_n(t) < s_n^2\Phi^{-1}\left(\frac{\alpha}{2}\right)\right\}$$

$$d_n = M_n^+ - M_n^-$$

Then

$$\hat{\sigma}_n^2 = \frac{\sqrt{n}d_n}{2\Phi^{-1}\left(1 - \frac{\alpha}{2}\right)} \xrightarrow{p} \sigma^2(\psi, F) \quad \text{as } n \to \infty$$

is proved, e.g., in Jurečková (1977).

If N_c is the stopping rule defined in (3.86) with $\hat{\sigma}_n^2$, then T_{N_c} is a risk-efficient M-estimator of θ.

Let now T_n be an L-estimator of θ, defined in (3.73), trimmed at α and $1-\alpha$, satisfying (3.74)–(3.77). Then $\sqrt{n}(T_n-\theta)$ is asymptotically normal with variance $\sigma^2(J,F)$ given in (3.78). Sen (1978) proposed the following estimator of $\sigma^2(J,F)$:

$$\hat{\sigma}_n^2 = \sum_{i=1}^{n-1}\sum_{j=1}^{n-1} c_{ni}c_{nj}[F(x\wedge y)-F(x)F(y)]J(F(x))J(F(y))dxdy$$

and proved that $\hat{\sigma}_n^2 \to \sigma^2(J,F)$ a.s. as $n\to\infty$.

Again, if N_c is the stopping rule defined in (3.86) with $\hat{\sigma}_n^2$, then T_{N_c} is a risk-efficient L-estimator of θ.

3.10   R-estimators

Let R_i be the rank of X_i among $X_1,\ldots,X_n$, $i=1,\ldots,n$, where $X_1,\ldots,X_n$ is a random sample from a population with a continuous distribution function. The rank R_i can be formally expressed as

$$R_i = \sum_{j=1}^{n} I[X_j \le X_i], \quad i=1,\ldots,n \tag{3.88}$$

and thus $R_i = nF_n(X_i)$, $i=1,\ldots,n$, where F_n is the empirical distribution function of $X_1,\ldots,X_n$. The ranks are invariant with respect to the class of monotone transformations of observations, and the tests based on ranks have many advantages: the most important among them is that the distribution of the test criterion under the hypothesis of randomness (i.e., if $X_1,\ldots,X_n$ are independent and identically distributed with a continuous distribution function) is independent of the distribution of observations.

Hodges and Lehmann (1963) proposed a class of estimators, called R-estimators, that are obtained by an inversion of the rank tests. The R-estimators can be defined for many models, practically for all where the rank tests make sense, and the test criterion is symmetric about a known center or has other suitable property under the null hypothesis. We shall describe the R-estimators of the center of symmetry of an (unknown) continuous distribution function, and the R-estimators in the linear regression model in the sequel.

Let $X_1,\ldots,X_n$ be independent random observations with continuous distribution function $F(x-\theta)$, symmetric about θ. The hypothesis

$$\boldsymbol{H}_0 : \ \theta = \theta_0$$

on the value of the center of symmetry is tested with the aid of the *signed rank test* (or one-sample rank test), based on the statistic

$$S_n(\theta_0) = \text{sign}(X_i - \theta_0)a_n(R_{ni}^+(\theta_0)) \tag{3.89}$$

where $R_{ni}^+(\theta_0)$ is the rank of $|X_i - \theta_0|$ among $|X_1 - \theta_0|, \ldots, |X_n - \theta_0|$ and $a_n(1) \le \ldots \le a_n(n)$ are given *scores*, generated by a nondecreasing score function $\varphi^+ : [0, 1) \mapsto \mathbb{R}^+$, $\varphi^+(0) = 0$, in the following way: $a_n(i) = \varphi^+\left(\frac{i}{n+1}\right)$, $i = 1, \ldots, n$. For example, the linear score function $\varphi^+(u) = u$, $0 \le u \le 1$ generates the *Wilcoxon one-sample test*. If θ_0 is the right center of symmetry, then are $\text{sign}(X_i - \theta_0)$ and $R_{ni}^+(\theta_0)$ stochastically independent, $i = 1, \ldots, n$, and $S_n(t)$ is a nonincreasing step function of t with probability 1 (Problem 3.10). This implies that $\mathbb{E}_{\theta_0} S_n(\theta_0) = 0$ and the distribution of $S_n(\theta_0)$ under $\boldsymbol{H}_0$ is symmetric around 0. As an estimator of θ_0 we propose the value of t which solves the equation $S_n(t) = 0$. Because $S_n(t)$ is discontinuous, such an equation may have no solution; then we define the R-estimator similarly as the M-estimator and put

$$T_n = \frac{1}{2}(T_n^- + T_n^+) \tag{3.90}$$

$$T_n^- = \sup\{t : S_n(t) > 0\}, \qquad T_n^+ = \inf\{t : S_n(t) < 0\}$$

T_n coincides with the sample median if $a_n(i) = 1$, $i = 1, \ldots, n$. The estimator, corresponding to the one-sample Wilcoxon test with the scores $a_n(i) = \frac{i}{n+1}$, $i = 1, \ldots, n$, is known as the *Hodges-Lehmann estimator*:

$$T_{nH} = \text{med}\left\{\frac{X_i + X_j}{2} : 1 \le i \le j \le n\right\} \tag{3.91}$$

We can compute this estimator by the function `hodges.lehmann` from Appendix A.

```
> hodges.lehmann(chem)
[1] 3.225
>
```

Other R-estimators should be computed by an iterative procedure.

Unlike the M-estimators, the R-estimators are not only translation, but also scale equivariant, i.e.,

$$T_n(X_1 + c, \ldots, X_n + c) = T_n(X_1, \ldots, X_n) + c, \ c \in \mathbb{R}$$

$$\tag{3.92}$$

$$T_n(cX_1, \ldots, cX_n) = cT_n(X_1, \ldots, X_n), \ c > 0$$

The distribution function of statistic $S_n(\theta)$ is discontinuous, even if $X_1, \ldots, X_n$ have a continuous distribution function $F(x - \theta)$. On the other hand, if θ is the actual center of symmetry, then the distribution function of statistic $S_n(\theta)$ does depend on F. If we denote

$$0 \le p_n = P_\theta(S_n(\theta) = 0) = P_0(S_n(0) = 0) < 1$$

then $\lim_{n\to\infty} p_n = 0$ and

$$\frac{1}{2}(1 - p_n) \le P_\theta(T_n < \theta) \le P_\theta(T_n \le \theta) \le \frac{1}{2}(1 + p_n) \qquad (3.93)$$

This means that if F is symmetric around zero, T_n is an asymptotically *median unbiased* estimator of θ.

Using (3.88) in statistic (3.89) with linear scores, we arrive at an alternative form of the Hodges-Lehmann estimator T_n as a solution of the equation

$$\int_{-\infty}^{\infty} [F_n(y) - F_n(2T_n - y)]dF_n(y) = 0 \qquad (3.94)$$

Similarly, the R-estimator generated by the score function φ^+ can be expressed as a solution of the equation

$$\int_{-\infty}^{\infty} \varphi\left(F_n(y) - F_n(2T_n - y)\right) dF_n(y) = 0 \qquad (3.95)$$

where $\varphi(u) = \text{sign}(u - \frac{1}{2})\varphi^+(2u - 1)$, $0 < u < 1$. Hence, the corresponding statistical functional is a solution of the equation

$$\int_{-\infty}^{\infty} \varphi[F(y) - F(2T(F) - y)] dF(y) \qquad (3.96)$$

$$= \int_0^1 \varphi\left[u - F(2T(F) - F^{-1}(u))\right] du = 0$$

The influence function of the R-estimator can be derived similarly as that of the L-estimator, and in case of symmetric F with an absolutely continuous density f it equals

$$IF(x, T, F) = \frac{\varphi(F(x))}{\int_{\mathbf{R}} \varphi(F(y))(-f'(y))dy} \qquad (3.97)$$

Example 3.5 *The breakdown point of Hodges-Lehmann estimator T_{nH}: the estimator can break down if at least half of sum $\frac{1}{2}(X_i + X_j)$ for $1 \le i \le j \le n$ is replaced. Assume a sample is corrupted by replacement of m outliers and $n + \binom{n}{2}$ is even, then the estimator T_{nH} breaks down for m satisfying*

$$n - m + \binom{n - m}{2} > \frac{1}{2}n + \frac{1}{2}\binom{n}{2}$$

Therefore
$$2m^2 - m(4n + 2) + n + n^2 > 0, \ 1 \le m \le n \qquad (3.98)$$

We look for the smallest integer m satisfying (3.98). Thus the breakdown point $\varepsilon_n^* = \frac{m_n^*}{n}$, where

$$m^* = \left\lceil \frac{2n + 1 - \sqrt{2n^2 + 2n + 1}}{2} + 1 \right\rceil$$

where $\lceil \cdot \rceil$ is the ceiling function, that is, $\lceil x \rceil$ is the smallest integer no smaller than x. Analogously, for $n + \binom{n}{2}$ odd

$$m^* = \left\lceil \frac{2n + 1 - \sqrt{2n^2 + 2n + 5}}{2} + 1 \right\rceil$$

Finally, $\lim_{n \to \infty} \varepsilon_n^*(T_{nH}) \doteq 0.293$.

Remark 3.4 *If $\psi(x) = c\varphi(F(x))$, $x \in \mathbb{R}$, then the influence function of the M-estimator generated by ψ coincides with the influence function of the R-estimator generated by φ.*

Jaeckel (1969) proposed an equivalent definition of the R-estimator, more convenient for the calculation. Consider the

$$n + \binom{n}{2} = \frac{n(n + 1)}{2}$$

averages $\frac{1}{2}(X_{n:j} + X_{n:k})$, $1 \le j \le k \le n$, including the cases $j = k$. Let $\varphi : (0, 1) \mapsto \mathbb{R}$ be a nondecreasing score function, skew-symmetric on $(0, 1)$ in the sense that

$$\varphi(1 - u) = -\varphi(u), \quad 0 < u < 1$$

and put

$$d_{in} = \varphi\left(\frac{i + 1}{2n + 1}\right) - \varphi\left(\frac{i}{2n + 1}\right) \quad i = 1, \ldots, n$$

Then define the weights c_{jk}, $1 \le j \le k \le n$ in the following way:

$$c_{jk} = \frac{d_{n-k+j}}{\sum_{i=1}^{n} id_i}; \qquad \sum_{1 \le j \le k \le n} c_{jk} = 1$$

The R-estimator is defined as the median of the discrete distribution that assigns the probability c_{jk} to each average $\frac{1}{2}(X_{n:j} + X_{n:k})$, $1 \le j \le k \le n$. If the score function φ is linear and hence $d_{n1} = \ldots = d_{nn} = \frac{1}{n}$, then the weights c_{jk} are all equal and the estimator is just the median of all averages, thus the Hodges-Lehmann estimator. However, Jaeckel's definition of the R-estimator is applicable to more general signed-rank tests, such as the one-sample van der Waerden test and others.

3.11   Examples

Example 3.6 *Assume that the following data are independent measurements of a physical entity θ: 46.34, 50.34, 48.35, 53.74, 52.06, 49.45, 49.90, 51.25, 49.38, 49.31, 50.62, 48.82, 46.90, 49.46, 51.17, 50.36, 52.18, 50.11, 52.49, 48.67.*

Otherwise, we have the measurements

$$X_i = \theta + e_i, \quad i = 1, \ldots, 20$$

and we want to determine the unknown value θ, assuming that the errors $e_1, \ldots, e_{20}$ are independent and identically distributed, symmetrically around zero. The first column of Table 3.2 provides the values of the estimates of θ and the scale characteristics from Table 3.1, based on $X_1, \ldots, X_{20}$.

Table 3.1 *Estimates of the location and the scale characteristics.*

Location	
mean	$\bar{X}_n$
median	$\tilde{X}_n$
5%-trimmed mean	$\bar{X}_{.05}$
10%-trimmed mean	$\bar{X}_{.10}$
5%-Winsorized mean	$\bar{W}_{.05}$
10%-Winsorized mean	$\bar{W}_{.05}$
Huber M-estimator	M_H
Hodges-Lehmann estimator	HL
Sen's weighted mean, $k_1 = [0.05n]$	S_{k_1}
Sen's weighted mean, $k_2 = [0.1n]$	S_{k_2}
midrange	$\mathcal{R}_m$
Scale	
standard deviation	S_n
inter-quartile range	R_I
median absolute deviation	MAD
Gini mean difference	G_n

We see that the values in column I are rather close to each other, and that the data seem to be roughly symmetric around 50.

Let us now consider what happens if some observations are slightly changed. The effects of some changes we can see in columns II-V of Table 3.2. Columns II and III illustrate the effects of a change of solely one observation, caused by a mistake in the decimal point: column II corresponds to the fact that the last value in the dataset, 48.67, was replaced by 486.7, while column III gives the result of a replacement of 48.67 by 4.867. These changes considerably affected the mean $\bar{X}_n$, the standard deviation S_n, and the midrange $\mathcal{R}_m$, which is in a correspondence with the theoretical conclusions that $\bar{X}_n$, S_n and $\mathcal{R}_m$ are highly non-robust.

Columns III and IV show the changes in the estimates when the last five observations in the dataset were replaced by the values 79.45, 76.80, 80.73, 76.10, 87.01, or by the values 1.92, 0.71, 1.26, 0.32, -1.71, respectively.

Table 3.2 *Effects of changes in the dataset on the estimates.*

estimator	I	II	III	IV	V
mean $\bar{X}_n$	50.04	71.95	47.85	57.36	37.48
median $\tilde{X}_n$	50.00	50.22	50.00	50.48	49.34
5%-trimmed mean $\bar{X}_{.05}$	50.05	50.33	49.92	56.32	38.75
10%-trimmed mean $\bar{X}_{.10}$	50.09	50.33	49.98	55.39	40.32
5%-Winsorized mean $\bar{W}_{.05}$	50.01	50.33	49.87	57.07	37.50
10%-Winsorized mean $\bar{W}_{.05}$	50.12	50.35	49.89	57.09	37.46
Huber M-estimator M_H	50.07	50.33	49.94	56.59	37.62
Hodges-Lehmann estimator HL	50.02	50.31	49.94	51.31	48.18
Sen's weighted mean $\mathcal{S}_{k_1}$	50.04	50.29	49.98	54.19	42.49
Sen's weighted mean $\mathcal{S}_{k_2}$	50.02	50.25	50.00	52.44	45.60
midrange $\mathcal{R}_m$	50.04	266.52	29.30	66.68	26.02
sample standard deviation S_n	1.82	97.64	10.28	13.67	21.97
inter-quartile range R_I	2.00	2.09	2.00	9.97	15.18
median absolute deviation MAD	1.18	0.98	1.18	1.62	1.86
Gini mean difference G_n	2.09	45.55	6.42	13.39	20.74

Example 3.7 *When we wish to obtain a picture of the behavior of an esti-mator under various models, we usually simulate the model and look at the resulting values of the estimator of interest. For example, 200 observations were simulated from the following probability distributions:*

- *Normal distribution $\mathcal{N}(0,1)$ and $\mathcal{N}(10,2)$ with the density*

$$f(x) = \frac{1}{\sigma\sqrt{2\pi}} e^{-\frac{(x-\mu)^2}{2\sigma^2}}, \ \mu = 0, \ 10, \ \sigma^2 = 1, \ 2, \ x \in \mathbb{R}$$

(symmetric and exponentially tailed distribution).

- *Exponential $Exp(1/5)$ distribution with the density*

$$f(x) = \frac{1}{5} e^{-\frac{x}{5}}, \ x \geq 0, \ f(x) = 0, \ x < 0$$

(skewed and exponentially tailed distribution).

- *Cauchy with the density*

$$f(x) = \frac{1}{\pi(1+x^2)}, \ x \in \mathbb{R}$$

(symmetric and heavy tailed distribution)

• *Pareto with the density*

$$f(x) = \frac{1}{(1+x)^2}, \ x \geq 1, \ f(x) = 0, \ x < 1$$

(skewed and heavy tailed distribution).

The values of various estimates under the above distributions are given in Table 3.3.

Table 3.3 *Values of estimates under various models.*

estimator	$\mathcal{N}(0,1)$	$\mathcal{N}(10,2)$	$Exp(1/5)$	Cauchy	Pareto
mean $\bar{X}_n$	0.06	9.92	4.49	1.77	12.19
median $\tilde{X}_n$	−0.01	9.73	2.92	−0.25	2.10
5%-trimmed mean $\bar{X}_{.05}$	0.05	9.89	4.01	−0.23	3.42
10%-trimmed mean $\bar{X}_{.10}$	0.04	9.88	3.75	−0.29	2.84
5%-Winsorized mean $\bar{W}_{.05}$	0.07	9.92	4.30	−0.10	4.17
10%-Winsorized mean $\bar{W}_{.05}$	0.05	9.91	4.05	−0.30	3.32
Huber M-estimator M_H	0.05	9.89	3.89	−0.29	2.87
Hodges-Lehmann est. HL	0.04	9.87	3.73	−0.24	2.73
Sen's weighted mean S_{k_1}	0.02	9.75	3.16	−0.22	2.24
Sen's weighted mean S_{k_2}	0.02	9.73	3.10	−0.21	2.17
midrange $\mathcal{R}_m$	0.05	10.37	11.85	146.94	525.34
sample standard deviation S_n	1.07	2.03	4.63	13.06	78.48
inter-quartile range R_I	1.26	3.05	5.53	2.45	2.83
median abs. deviation MAD	0.63	1.58	2.02	1.24	0.93
Gini mean difference G_n	1.18	2.30	4.78	7.18	20.23

Example 3.8 *In case of an M-estimator with redescending ψ-function, the solution can depend on the starting value or even the procedure does not have to converge. To illustrate this fact we apply the function* `skipped.mean` *on the following artificial data set.*

```
> set.seed(256)
> y <- c(rnorm(25), rnorm(5)-5,rnorm(10)+5)
> skipped.mean(y, initmu = median(y), iters = F)
$mu
[1] 0.4452402

$s
[1] 2.154085

$Niter
[1] 13
```

```
$Converged
[1] TRUE

> skipped.mean(y, initmu = 0, iters = F)
$mu
[1] -0.05916498

$s
[1] 2.154085

$Niter
[1] 14

$Converged
[1] TRUE

> skipped.mean(y, initmu = mean(y), iters = T)
$mu
[1] 1.01604

$s
[1] 2.768372

$Niter
[1] 300

$Converged
[1] FALSE

>
```

3.12   Problems and complements

3.1 Let $X_1, \ldots, X_n$ be a sample from the distribution with the density

$$f(x) = \begin{cases} 1 & \text{if} \quad |x| \le \frac{1}{4} \\ \frac{1}{32|x|^3} & \text{if} \quad |x| > \frac{1}{4} \end{cases}$$

and $\bar{X} = \frac{1}{n}\sum_{i=1}^{n} X_i$ be the sample mean. Then var $\bar{X} = \infty$.

3.2 The α-*interquantile range* $(0 < \alpha < 1)$:

$$S_\alpha = F^{-1}(1 - \alpha) - F^{-1}(\alpha).$$

The influence function of S_α equals

$$
IF(x; F, S_\alpha) = \begin{cases} \frac{1-\alpha}{f(a_2)} - \frac{\alpha}{a_1} & \dots & x < a_1 \\ -\alpha \left[\frac{1}{f(a_1)} + \frac{1}{f(a_2)} \right] & \dots & a_1 < x < a_2 \\ \frac{1-\alpha}{f(a_2)} - \frac{\alpha}{f(a_1)} & \dots & x > a_2 \end{cases}
$$

where $a_1 = F^{-1}(\alpha)$ and $a_2 = F^{-1}(1-\alpha)$ and $f(x) = \frac{dF(x)}{dx}$; the derivative should exist in neighborhoods of a_1 and a_2.

3.3 The *symmetrized α-interquantile range* $(0 < \alpha < 1)$ (Collins (2000)):

$$
\tilde{S}_\alpha(F) = S_\alpha(\tilde{F}) = \tilde{F}^{-1}(1-\alpha) - \tilde{F}^{-1}(\alpha)
$$

where

$$
\tilde{F}(x) = \tfrac{1}{2}\left\{ F(x) + 1 - F\left[2F^{-1}(\tfrac{1}{2}) - x \right] \right\} \quad \text{for} \quad F \quad \text{continuous}
$$

$\tilde{S}_{\frac{1}{4}}$ coincides with MAD. Calculate the influence function of $\tilde{S}_\alpha$.

3.4 Let $X_1, \dots, X_n$ be an independent sample from a population with density $f(x - \theta)$ and let $T(X_1, \dots, X_n)$ be a translation equivariant estimator of θ, then T_n has a continuous distribution function.

  Hint: $T(x_1, \dots, x_n) = t$ if and only if $x_1 = t - T(0, x_2 - x_1, \dots, x_n - x_1)$.

  Hence, given $X_2 - X_1, \dots, X_n - X_1 = (y_2, \dots, y_n)$ and $t \in \mathbb{R}$, there is exactly one point $\mathbf{x}$ for which $T(\mathbf{x}) = t$. Hence, $P\{T(\mathbf{X}) = t | X_2 - X_1 = y_2, \dots, X_n - X_1 = y_n\} = 0$ for every $(y_2, \dots, y_n)$ and t, thus $P\{T(\mathbf{X}) = t\} = 0 \ \forall t$.

3.5 Let $X_1, \dots, X_n$ be independent observations with distribution function $F(x - \theta)$, and let $T_n = \sum_{i=1}^n c_{ni} X_{n:i}$ be an L-estimator of θ, then:

(a) If $\sum_{i=1}^n c_{ni} = 1$, then T is translation equivariant.

(b) If F is symmetric about zero, $\sum_{i=1}^n c_{ni} = 1$ and $c_{ni} = c_{n,n-i+1}$, $i = 1, \dots, n$, then the distribution of T_n is symmetric about θ.

3.6 Tukey (1960) proposed the model of the normal distribution with variance 1 contaminated by the normal distribution with variance $\tau^2 > 1$, i.e., that of the distribution function F of the form

$$
F(x) = (1 - \varepsilon)\Phi(x) + \varepsilon\Phi\left(\frac{x}{\tau}\right) \tag{3.99}
$$

where Φ is the standard normal distribution function. Compare the asymptotic variances of the sample mean and the sample variance under (3.99).

3.7 Let $X_1, \dots, X_n$ be a sample from the Cauchy distribution $C(\xi, \sigma)$ with the density

$$
f(x) = \frac{\sigma}{\pi} \frac{1}{\sigma^2 + (x - \xi)^2}
$$

Then the distribution of $\bar{X}_n$ is again $C(\xi, \sigma)$

3.8 Let X, $-\frac{\pi}{2} \leq X \leq \frac{\pi}{2}$, be a random angle with the uniform distribution on the unit circle. Then tg X has the Cauchy distribution $C(0,1)$.

3.9 Consider the equation

$$\sum_{i=1}^{n} \psi_C(X_i - \theta) = 0$$

where ψ_C is the Cauchy likelihood function (3.18). Denote K_n as the number of its roots. If $X_1, \ldots, X_n$ are independent, identically distributed with the Cauchy $C(0,1)$ distribution, then $K_n - 1$ has asymptotically Poisson distribution with parameter $\frac{1}{\pi}$, as $n \to \infty$. (See Barnett (1966) or Reeds (1985)).

3.10 (a) If X is a random variable with a continuous distribution function, symmetric about zero, then sign X and $|X|$ are stochastically independent. (b) Prove that the linear signed rank statistic (3.89) is a nondecreasing step function of θ with probability 1, provided the score function φ^+ is nondecreasing on $(0,1)$ and $\varphi^+(0) = 0$. (See van Eeden (1972) or Jurečková and Sen (1996), Section 6.4.)

3.11 Generate samples of different distribution and apply the described methods to these data.

3.12 Apply the described methods to the variable `logst` of the dataset `CYGOB1` from package `HSAUR`.

3.13 Write an R procedure that computes the M-estimator with Andrews sinus ψ-function and skipped median ψ-function.

3.14 Apply the functions `huber2`, `cauchy`, `hampel`, `skipped.mean` and `biweight` from Appendix A with different values of the arguments on the dataset of Example 3.8.

Chapter 4

Linear model

4.1 Introduction

Consider the linear regression model

$$Y_i = \boldsymbol{x}_i' \boldsymbol{\beta} + U_i, \ i = 1, \ldots, n \qquad (4.1)$$

with observations $Y_1, \ldots, Y_n$, unknown and unobservable parameter $\boldsymbol{\beta} \in \mathbb{R}^p$, where $\boldsymbol{x}_i \in \mathbb{R}^p$, $i = 1, \ldots, n$ are either given deterministic vectors or observable random vectors (*regressors*) and $U_1, \ldots, U_n$ are independent errors with a joint distribution function F. Often we consider the model in which the first component β_1 of $\boldsymbol{\beta}$ is an intercept: it means that $x_{i1} = 1$, $i = 1, \ldots, n$. Distribution function F is generally unknown; we only assume that it belongs to some family $\mathcal{F}$ of distribution functions.

Denoting

$$\boldsymbol{Y} = (Y_1, \ldots, Y_n)'$$

$$\boldsymbol{X} = \boldsymbol{X}_n = \begin{pmatrix} \boldsymbol{x}_1' \\ \vdots \\ \boldsymbol{x}_n' \end{pmatrix}$$

$$\boldsymbol{U} = (U_1, \ldots, U_n)'$$

we can rewrite (4.1) in the matrix form

$$\boldsymbol{Y} = \boldsymbol{X}\boldsymbol{\beta} + \boldsymbol{U} \qquad (4.2)$$

The most popular estimator of $\boldsymbol{\beta}$ is the classical *least squares estimator* (LSE) $\widehat{\boldsymbol{\beta}}$. If $\boldsymbol{X}$ is of rank p, then $\widehat{\boldsymbol{\beta}}$ is equal to

$$\widehat{\boldsymbol{\beta}} = (\boldsymbol{X}'\boldsymbol{X})^{-1}\boldsymbol{X}'\boldsymbol{Y} \qquad (4.3)$$

As it follows from the Gauss-Markov theorem, $\widehat{\boldsymbol{\beta}}$ is the best linear unbiased estimator of $\boldsymbol{\beta}$, provided the errors $U_1, \ldots, U_n$ have a finite second moment. Moreover, $\widehat{\boldsymbol{\beta}}$ is the maximum likelihood estimator of $\boldsymbol{\beta}$ if $U_1, \ldots, U_n$ are normally distributed.

The least squares estimator $\widehat{\beta}$ is an extension of the sample mean to the linear regression model. Then, naturally, it has similar properties: it is highly non-robust and sensitive to outliers and to the gross errors in Y_i, and to the deviations from the normal distribution of errors. It fails if the distribution of the U_i is heavy-tailed. But above all this, $\widehat{\beta}$ is heavily affected by the regression matrix $\boldsymbol{X}$, namely it is sensitive to the outliers among its elements. Violating some conditions in the linear models can have more serious consequences than in the location model. This can have a serious impact in econometric, but also in many other applications. Hence, we must look for robust alternatives to the classical procedures in linear models.

Example 4.1 *Figure 4.1 illustrates an effect of an outlier in the x-direction (leverage point) on the least squares estimator.*

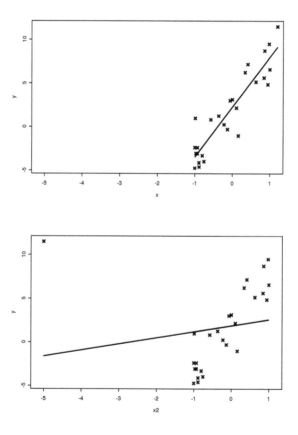

Figure 4.1 *Data with 27 points and the corresponding least squares regression line (top) and the sensitivity of least squares regression to an outlier in the x-direction (bottom).*

Before we start describing the robust statistical procedures, we shall try to illustrate how seriously the outliers in $\boldsymbol{X}$ can affect the performance of the estimator $\widehat{\boldsymbol{\beta}}$.

4.2   Least squares method

If we estimate $\boldsymbol{\beta}$ by the least squares method, then the set $\widehat{\boldsymbol{Y}} = \boldsymbol{X}\widehat{\boldsymbol{\beta}}$ is a hyperplane passing through the points $(\boldsymbol{x}_i, \hat{Y}_i)$, $i = 1, \ldots, n$, where

$$\hat{Y}_i = \boldsymbol{x}_i'\widehat{\boldsymbol{\beta}} = \boldsymbol{h}_i'\boldsymbol{Y}, \ i = 1, \ldots, n$$

and $\boldsymbol{h}_i'$ is the ith row of the project (hat) matrix $\widehat{\boldsymbol{H}} = \boldsymbol{X}\left(\boldsymbol{X}'\boldsymbol{X}\right)^{-1}\boldsymbol{X}'$. Hence, $\widehat{\boldsymbol{Y}} = \widehat{\boldsymbol{H}}\boldsymbol{Y}$ is the projection of vector $\boldsymbol{Y}$ in the space spanned by the columns of matrix $\boldsymbol{X}$.

Because $\widehat{\boldsymbol{H}}$ is the project matrix, then $\boldsymbol{h}_i'\boldsymbol{h}_j = h_{ij}$, $i, j = 1, \ldots, n$, and thus

$$0 \le \sum_{k \ne i} h_{ik}^2 = h_{ii}(1 - h_{ii}) \Longrightarrow 0 \le h_{ii} \le 1, \ i = 1, \ldots, n$$

$$(4.4)$$

$$\Longrightarrow |h_{ij}| \le \|\boldsymbol{h}_i\| \cdot \|\boldsymbol{h}_j\| = (h_{ii}h_{jj})^{\frac{1}{2}} \le 1, \ i, j = 1, \ldots, n$$

The matrix $\widehat{\boldsymbol{H}}$ is of order $n \times n$ and of rank p; its diagonal elements lie in the interval $0 \le h_{ii} \le 1$, $i = 1, \ldots, n$ and trace$\widehat{\boldsymbol{H}} = \sum_{i=1}^{n} h_{ii} = p$.

In the extreme situation we can imagine that $h_{ii} = 1$ for some i; then

$$1 = h_{ii} = \|\boldsymbol{h}_i\|^2 = \sum_{k=1}^{n} h_{ik}^2 = 1 + \sum_{k \ne i} h_{ij}^2$$

$$\Longrightarrow h_{ij} = 0 \ \text{ for } \ j \ne i$$

which means that
$$\hat{Y}_i = \boldsymbol{x}_i'\widehat{\boldsymbol{\beta}} = \boldsymbol{h}_i'\boldsymbol{Y} = h_{ii}Y_i = Y_i$$

and the regression hyperplane passes through $(\boldsymbol{x}_i, Y_i)$, regardless of the values of other observations. The value $h_{ii} = 1$ is an extreme case, but it illustrates that a high value of the diagonal element h_{ii} causes the regression hyperplane to pass near to the point $(\boldsymbol{x}_i, Y_i)$. This point is called a *leverage point* of the dataset. There are different opinions about which value h_{ii} can be considered as high. Huber (1981, p. 162) considers $\boldsymbol{x}_i$ as a leverage point if $h_{ii} > .5$. It is well known that if $\mathbb{E}U_i = 0$ and $0 < \sigma^2 = \mathbb{E}U_i^2 < \infty$, $i = 1, \ldots, n$, then

$$\lim_{n \to \infty} \max_{1 \le i \le n} h_{ii} = 0$$

is a necessary and sufficient condition for the convergence

$$\mathbb{E}_\beta \|\widehat{\boldsymbol{\beta}}_n - \boldsymbol{\beta}\|^2 \to 0$$

$$\mathcal{L}\left\{ (\boldsymbol{X}'\boldsymbol{X})^{-\frac{1}{2}} (\widehat{\boldsymbol{\beta}}_n - \boldsymbol{\beta}) \right\} \to \mathcal{N}\left(\boldsymbol{0}, \sigma^2 \boldsymbol{I}_p\right)$$

as $n \to \infty$, where $\boldsymbol{I}_p$ is the identity matrix of order p (see, e.g., Huber (1981)).

It is intuitively clear that large values of the residuals $|\widehat{Y}_i - Y_i|$ are caused by a large maximal diagonal element of $\widehat{\boldsymbol{H}}$. Consider this relation in more detail. Assume that the distribution function F has nondegenerate tails, i.e., $0 < F(x) < 1$, $x \in \mathbb{R}$; moreover, assume that it is symmetric around zero, i.e., $F(x) + F(-x) = 1$, $x \in \mathbb{R}$, for the sake of simplicity. One possible characteristic of the tail-behavior of estimator $\widehat{\boldsymbol{\beta}}$ is the following measure:

$$B(a, \widehat{\boldsymbol{\beta}}) = \frac{-\log \mathbb{P}_\beta \left(\max_i |\boldsymbol{x}_i'(\widehat{\boldsymbol{\beta}} - \boldsymbol{\beta})| > a \right)}{-\log(1 - F(a))} \tag{4.5}$$

We naturally expect that

$$\lim_{a \to \infty} \mathbb{P}_\beta \left(\max_i |\boldsymbol{x}_i'(\widehat{\boldsymbol{\beta}} - \boldsymbol{\beta})| > a \right) = 0 \tag{4.6}$$

and we are interested in how fast this convergence can be, and under which conditions it is faster. The faster convergence leads to larger values of (4.5) under large a, denote

$$\tilde{h} = \max_{1 \le i \le n} h_{ii}, \quad h_{ii} = \boldsymbol{x}_i'(\boldsymbol{X}'\boldsymbol{X})^{-1}\boldsymbol{x}_i, \quad i = 1, \ldots, n \tag{4.7}$$

The influence of $\tilde{h}$ on the limit behavior of $B(a, \widehat{\boldsymbol{\beta}})$ is described in the following theorem:

Theorem 4.1 *Let $\widehat{\boldsymbol{\beta}}$ be the least squares estimator of $\boldsymbol{\beta}$ in model (4.1) with a nonrandom matrix $\boldsymbol{X}$.*

(i) If F has exponential tails, i.e.,

$$\lim_{a \to \infty} \frac{-\log(1 - F(a))}{ba} = 1, \quad b > 0, \quad then$$

$$\frac{1}{\sqrt{\tilde{h}}} \le \underline{\lim}_{a \to \infty} B(a, \widehat{\boldsymbol{\beta}}) \le \overline{\lim}_{a \to \infty} B(a, \widehat{\boldsymbol{\beta}}) \le \frac{1}{\tilde{h}} \tag{4.8}$$

(ii) If F has exponential tails with exponent r, i.e.,

$$\lim_{a \to \infty} \frac{-\log(1 - F(a))}{ba^r} = 1, \quad b > 0 \quad and \quad r \in (1, 2]$$

then

$$\tilde{h}^{-r+1} \le \underline{\lim}_{a \to \infty} B(a, \widehat{\boldsymbol{\beta}}) \le \overline{\lim}_{a \to \infty} B(a, \widehat{\boldsymbol{\beta}}) \le \tilde{h}^{-r} \tag{4.9}$$

(iii) If F is a normal distribution function, then

$$\lim_{a \to \infty} B(a, \widehat{\boldsymbol{\beta}}) = \frac{1}{\tilde{h}} \tag{4.10}$$

(iv)   If F is heavy-tailed, i.e.,

$$\lim_{a \to \infty} \frac{-\log(1 - F(a))}{m \log a} = 1, \quad m > 0$$

then

$$\lim_{a \to \infty} B(a, \widehat{\boldsymbol{\beta}}) = 1 \tag{4.11}$$

Theorem 4.1 shows that if the maximal diagonal element $\tilde{h}$ of matrix $\widehat{\boldsymbol{H}}$ is large, then the probability $\mathbb{P}_{\boldsymbol{\beta}}(\max_i |\boldsymbol{x}_i'(\widehat{\boldsymbol{\beta}} - \boldsymbol{\beta}| > a)$ decreases slowly to 0 with increasing a; this is the case even when F is normal and the number n of observations is large. The upper bound of $B(a, \widehat{\boldsymbol{\beta}})$ under normal F is

$$\overline{\lim}_{a \to \infty} B(a, \widehat{\boldsymbol{\beta}}) \le \frac{n}{p} \tag{4.12}$$

with the equality under a balanced design with the diagonal $h_{ii} = \frac{p}{n}$, $i = 1, \dots, n$.

PROOF OF THEOREM 4.1. Let us assume, without loss of generality, that $\tilde{h} = h_{11}$. Because $0 < \tilde{h} \le 1$ and $\hat{Y}_i = \boldsymbol{x}_i'\widehat{\boldsymbol{\beta}} = \boldsymbol{h}_i'\boldsymbol{Y}$, we can write

$$\mathbb{P}_{\boldsymbol{\beta}}\left(\max_i \left|\boldsymbol{x}_i'(\widehat{\boldsymbol{\beta}} - \boldsymbol{\beta}\right| > a\right)$$

$$= \mathbb{P}_0(\max_i |\boldsymbol{h}_i'\boldsymbol{Y}| > a) \ge \mathbb{P}_0(\boldsymbol{h}_1'\boldsymbol{Y} > a)$$

$$\ge \mathbb{P}_0(\tilde{h}Y_1 > a, h_{12}Y_2 \ge 0, \dots, h_{1n}Y_n \ge 0)$$

$$\ge \mathbb{P}_0(Y_1 > a/\tilde{h})\left(\tfrac{1}{2}\right)^{n-1} = (1 - F(a/\tilde{h}))\left(\tfrac{1}{2}\right)^{n-1}$$

This implies that

$$\overline{\lim}_{a \to \infty} B(a, \widehat{\boldsymbol{\beta}}) \le \overline{\lim}_{a \to \infty} \frac{-\log(1 - F(a/\tilde{h}))}{-\log(1 - F(a))} \tag{4.13}$$

If F has exponential tails with index r, then it further follows from (4.13) that

$$\overline{\lim}_{a \to \infty} B(a, \widehat{\boldsymbol{\beta}}) \le \overline{\lim}_{a \to \infty} \frac{b(a/\tilde{h})^r}{ba^r} = \tilde{h}^{-r}$$

which gives the upper bounds in (i) and (ii). For a heavy-tailed F, it follows from (4.13) that

$$\overline{\lim}_{a \to \infty} B(a, \widehat{\boldsymbol{\beta}}) \le \overline{\lim}_{a \to \infty} \frac{m \log(a/\tilde{h})}{m \log a} = 1$$

and it gives (iv) because $\widehat{\boldsymbol{\beta}}$ has both positive and negative residuals and $\lim_{a \to \infty} B(a, \widehat{\boldsymbol{\beta}}) \geq 1$.

It remains to verify the lower bounds in (ii) and (iii). If F has exponential tails with exponent r, $1 < r \leq 2$, then, using the Markov inequality, we can write for any $\varepsilon \in (0, 1)$ that

$$\mathbb{P}_{\boldsymbol{\beta}}(\max_i |\boldsymbol{x}_i'(\widehat{\boldsymbol{\beta}} - \boldsymbol{\beta})| > a) \leq \frac{\mathbb{E}_0[\exp\{(1 - \varepsilon)b\tilde{h}^{1-r}(\max_i |\hat{Y}_i|^r)\}]}{\exp\{(1 - \varepsilon)b\tilde{h}^{1-r}a^r\}}$$

Hence, if we can verify that

$$\mathbb{E}_0[\exp\{(1 - \varepsilon)b\tilde{h}^{1-r}(\max_i |\hat{Y}_i|)^r\}] \leq C_r < \infty \qquad (4.14)$$

we can claim that

$$-\log \mathbb{P}_0(\max_i |\hat{Y}_i| > a) \geq -\log C_r + (1 - \varepsilon)b\tilde{h}^{1-r}a^r$$

and this would give the lower bound in (ii), and in fact also the lower bound for the normal distribution in (iii). Thus, it remains to prove that the expected value in (4.14) is finite. Denote $\|\boldsymbol{x}\|_s = (\sum_{i=1}^n |x_i|^s)^{1/s}$, $s > 0$ and put $s = \frac{r}{r-1}$ (> 2). Then, regarding the relation $\sum_{k=1}^n h_{ik}^2 = h_{ii}$, we conclude

$$(\max_i |\hat{Y}_i|)^r = \max_i |\boldsymbol{h}_i'\boldsymbol{Y}|^r \leq \max_i (\|\boldsymbol{h}_i\|_s \|\boldsymbol{Y}\|_r)^r$$

$$\leq \max_i (\sum_{k=1}^n h_{ik}^2)^{r/s} \sum_{k=1}^n |Y_k|^r \leq \tilde{h}^{r-1} \sum_{k=1}^n |Y_k|^r$$

and hence

$$\mathbb{E}_0 \exp\{(1 - \varepsilon)b\tilde{h}^{1-r}(\max_i |\hat{Y}_i|^r\} \leq \mathbb{E}_0 \exp\{(1 - \varepsilon)b \sum_{k=1}^n |Y_k|^r\}$$

$$\leq (\mathbb{E}_0 \exp\{(1 - \varepsilon)b|Y_1|^r\})^n$$

For exponentially tailed F with exponent r, there exists $K > 0$ such that

$1 - F(x) \le \exp\{-(1 - \frac{\varepsilon}{2}bx^r\} = C_K$ for $x > K$. Integrating by parts, we obtain

$$0 < \mathbb{E}_0[\exp\{(1 - \varepsilon)b|Y_1|^r\}] = -2\int_0^\infty \exp\{(1 - \varepsilon)by^r\}d(1 - F(y))$$

$$\le 2\int_0^K \exp\{(1 - \varepsilon)by^r\}dF(y) + 2\exp\{(1 - \varepsilon)bK^r\}(1 - F(K))$$

$$+2\int_K^\infty r(1 - \varepsilon)by^{r-1}(1 - F(y))\exp\{(1 - \varepsilon)by^r\}dy \qquad (4.15)$$

$$\le 2\int_0^K \exp\{(1 - \varepsilon)by^r\}dF(y) + 2(1 - F(K))\exp\{(1 - \varepsilon)bK^2\}$$

$$+2\int_K^\infty r(1 - \varepsilon)by^{r-1}\exp\{-\frac{\varepsilon}{2}by^r\}dy \le C_\varepsilon < \infty$$

So we have proved (4.14) for $1 < r \le 2$. If $r = 1$, we proceed as follows: (4.4) implies that $|h_{ij}| \le \sqrt{h_{ii}}$, $i, j = 1, \ldots, n$; thus

$$\max_i |\hat{Y}_i| = \max_i |\boldsymbol{h}_i'\boldsymbol{Y}| = \max_i |\sum_{j=1}^n h_{ij}Y_j|$$

$$\le \max_{ij} |h_{ij}| \sum_{j=1}^n |Y_j| \le \max_i |\sum_{j=1}^n h_{ij}Y_j| \le \tilde{h}^{\frac{1}{2}}\sum_{j=1}^n |Y_j|$$

Using the Markov inequality, we obtain

$$\mathbb{P}_0(\max_i |\hat{Y}_i| > a) \le \frac{\mathbb{E}_0\exp\{(1 - \varepsilon)b\tilde{h}^{-\frac{1}{2}}\max_i |\hat{Y}_i|\}}{\exp\{(1 - \varepsilon)b\tilde{h}^{-\frac{1}{2}}a\}}$$

$$\le \frac{(\mathbb{E}_0\exp\{(1 - \varepsilon)b|Y_1|\})^n}{\exp\{(1 - \varepsilon)b\tilde{h}^{-\frac{1}{2}}a\}}$$

and it further follows from (4.15) that $\mathbb{E}_0\exp\{(1 - \varepsilon)b|Y_1|\} < \infty$; this gives the lower bound in (i).

If F is the normal distribution function of $\mathcal{N}(0, \sigma^2)$, then $\hat{\boldsymbol{Y}} - \boldsymbol{X}\boldsymbol{\beta}$ has n-dimensional normal distribution $\mathcal{N}_n\left(\boldsymbol{0}, \sigma^2\widehat{\boldsymbol{H}}\right)$, hence

$$\mathbb{P}_0(\max_i |\hat{Y}_i| > a) \ge \mathbb{P}_0(\boldsymbol{h}_1'\boldsymbol{Y} > a) = 1 - \Phi(a\sigma^{-1}\tilde{h}^{-\frac{1}{2}})$$

and $\overline{\lim}_{a\to\infty}B(a, \widehat{\boldsymbol{\beta}}) \le \tilde{h}^{-1}$. $\square$

The proposition (iii) of Theorem 4.1 shows that the performance of the LSE can be poor even under the normal distribution of errors, provided the design is fixed and contains leverage points leading to large $\tilde{h}$. The rate of convergence in (4.6) does not improve even if the number of observations increases. In the extreme case of $\tilde{h} = 1$, the convergence (4.6) to zero is equally slow for arbitrarily large number n of observations as if there is only one observation (just the leverage one).

On the other hand, if the design is balanced with diagonal elements $h_{ii} = \frac{p}{n}$, $i = 1, \ldots, n$, then $\lim_{a \to \infty} B(a, \widehat{\boldsymbol{\beta}}_n) = \frac{n}{p}$, hence the rate of convergence in (4.6) improves with n.

Up to now, we have considered a fixed number n of observations. The situation can change if $n \to \infty$ and $a = a_n \to \infty$ at an appropriate rate. An interesting choice of the sequence $\{a_n\}$ is the population analogue of the extreme error among $U_1, \ldots, U_n$, i.e., $a = a_n = F^{-1}\left(1 - \frac{1}{n}\right)$. For the normal distribution $\mathcal{N}(0, \sigma^2)$, this population extreme is approximately

$$a_n = \sigma \Phi^{-1}\left(1 - \frac{1}{n}\right) \approx \sigma\sqrt{2\log n} \tag{4.16}$$

Namely, for the normal distribution of errors, we shall derive the lower and upper bounds of $B(a_n, \widehat{\boldsymbol{\beta}}_n)$ under this choice of $\{a_n\}$. These bounds are more optimistic for the least squares estimator, because they both improve with increasing n. This is true for both fixed and random designs, and in the latter case even when the $\boldsymbol{x}_i$ are random vectors with a heavy-tailed distribution G, $i = 1, \ldots, n$.

Theorem 4.2 *Consider the linear regression model*

$$Y_i = \boldsymbol{x}_i'\boldsymbol{\beta} + e_i, \quad i = 1, \ldots, n \tag{4.17}$$

with the observations Y_i, $i = 1, \ldots, n$ and with independent errors $e_1, \ldots, e_n$, normally distributed as $\mathcal{N}(0, \sigma^2)$. Assume that the matrix $\boldsymbol{X}_n = [\boldsymbol{x}_1', \ldots, \boldsymbol{x}_n']'$ is either fixed and of rank p for $n \geq n_0$ or that $\boldsymbol{x}_1, \ldots, \boldsymbol{x}_n$ are independent p-dimensional random vectors, independent of the errors, identically distributed with distribution function G, then

$$\overline{\lim}_{n \to \infty} \frac{B(\sigma\sqrt{2\log n}, \widehat{\boldsymbol{\beta}}_n)}{\left(\frac{n}{p}\right)^2 + \frac{(n-1)\log 2}{\log n}} \leq 1 \tag{4.18}$$

and

$$\underline{\lim}_{n \to \infty} \frac{B\left(\sigma\sqrt{2\log n}, \widehat{\boldsymbol{\beta}}_n\right)}{\frac{n}{2}\left(1 - \frac{1+\log 2}{\log n} - \frac{\log\log n}{\log n}\right)} \geq 1 \tag{4.19}$$

Remark 4.1 *The bounds (4.18) and (4.19) can be rewritten in the form of the following asymptotic inequalities that are true for the LSE under normal*

F, as $n \to \infty$ and for any $\varepsilon > 0$:

$$\left(\frac{n}{p}\right)^2 \geq \frac{-\log \, \mathbb{P}_0 \left(\max_i \left(\boldsymbol{x}_i' \widehat{\boldsymbol{\beta}}_n \right) \geq (\sigma \sqrt{2 \log \, n} \right)}{\log n} \tag{4.20}$$

$$\geq \frac{n}{2} \left\{ 1 - \frac{1 + \log 2}{\log n} - \frac{\log \log n}{\log n} \right\} \geq \frac{n(1 - \varepsilon)}{2}$$

PROOF OF THEOREM 4.2. Let us first consider the upper bound. Note that $\tilde{h}_n \geq \frac{p}{n}$ because trace $\widehat{\boldsymbol{H}}_n = p$. For each j such that $h_{jj} > 0$ we can write

$$\mathbb{P}_0 \left(\max_{1 \leq i \leq n} (\boldsymbol{x}_i' \widehat{\boldsymbol{\beta}}_n) \geq a \middle| (\boldsymbol{x}_1, \ldots, \boldsymbol{x}_n) \right) \tag{4.21}$$

$$= \mathbb{P}_0 \left(\max_i (\boldsymbol{h}_i' \boldsymbol{Y}) \geq a \middle| (\boldsymbol{x}_1, \ldots, \boldsymbol{x}_n) \right) \geq \mathbb{P}_0 \left(\boldsymbol{h}_j' \boldsymbol{Y} \geq a \middle| (\boldsymbol{x}_1, \ldots, \boldsymbol{x}_n) \right)$$

$$\geq \mathbb{P}_0 \left(h_{jj} Y_j \geq a, h_{1k} Y_k \geq 0, \, k \neq j \middle| (\boldsymbol{x}_1, \ldots, \boldsymbol{x}_n) \right)$$

$$\geq \mathbb{P}_0 \left(Y_j \geq \frac{a}{h_{jj}} \middle| (\boldsymbol{x}_1, \ldots, \boldsymbol{x}_n) \right) \left(\frac{1}{2}\right)^{n-1} = \left(1 - F \left(\frac{a}{h_{jj}} \right) \right) \left(\frac{1}{2}\right)^{n-1}$$

This holds for each j such that $h_{jj} > 0$; hence also

$$\mathbb{P}_0 \left(\max_{1 \leq i \leq n} (\boldsymbol{x}_i' \widehat{\boldsymbol{\beta}}_n) \geq a \middle| (\boldsymbol{x}_1, \ldots, \boldsymbol{x}_n) \right) \geq \left(1 - F \left(\frac{a}{\tilde{h}} \right) \right) \left(\frac{1}{2}\right)^{n-1}$$

Because $(-\log)$ is a convex function, we may apply the Jenssen inequality and (4.21) then implies

$$-\log \, \mathbb{E}_G \mathbb{P}_0 \left(\max_i (\boldsymbol{x}_i' \widehat{\boldsymbol{\beta}}_n) \geq a \middle| (\boldsymbol{x}_1, \ldots, \boldsymbol{x}_n) \right)$$

$$\leq (n - 1) \log \, 2 - \log \left\{ \mathbb{E}_G \left(1 - F \left(\frac{a}{\tilde{h}} \right) \right) \right\} \tag{4.22}$$

$$\leq (n - 1) \log 2 + \mathbb{E}_G \left\{ -\log \left(1 - F \left(\frac{a}{\tilde{h}} \right) \right) \right\}$$

If F is of exponential type (4.9) with $1 \leq r \leq 2$, then

$$a = a_n = \left(b^{-1} \log n \right)^{\frac{1}{r}} \to \infty \, \text{ as } n \to \infty$$

and (4.22) gives the following asymptotic inequality, as $n \to \infty$, for any G:

$$B(a_n, \widehat{\boldsymbol{\beta}}_n) \leq \mathbb{E}_G \left(\frac{1}{\tilde{h}_n} \right)^r + \frac{(n - 1) \log 2}{\log n} \leq \left(\frac{n}{p} \right)^r + \frac{(n - 1) \log 2}{\log n}$$

A more precise form of the above inequality is

$$\overline{\lim}_{n \to \infty} \frac{B(a_n, \widehat{\boldsymbol{\beta}}_n)}{\left(\frac{n}{p} \right)^r + \frac{(n-1) \log 2}{\log n}} \leq 1$$

and this gives (4.18).

The lower bound: Since $\mathcal{N}(0,\sigma^2)$ has exponential tails,

$$\lim_{a\to\infty} \frac{-\log(1-\Phi(a/\sigma))}{a^2/2\sigma^2} = 1 \tag{4.23}$$

and we can write

$$
\begin{aligned}
\mathbb{P}_0\left(\max_{1\le i\le n}|\boldsymbol{x}_i'\widehat{\boldsymbol{\beta}}_n| \ge a \,\Big|\, (\boldsymbol{x}_1,\ldots,\boldsymbol{x}_n)\right)
&= \mathbb{P}_0\left(\max_{1\le i\le n}|\boldsymbol{h}_i'\boldsymbol{Y}| \ge a \,\Big|\, (\boldsymbol{x}_1,\ldots,\boldsymbol{x}_n)\right) \\
\le \mathbb{P}_0\left(\tilde{h}^{\frac{1}{2}}\|\boldsymbol{Y}\| \ge a \,\Big|\, (\boldsymbol{x}_1,\ldots,\boldsymbol{x}_n)\right)
&= \mathbb{P}_0\left(\frac{\|\boldsymbol{Y}\|^2}{\sigma^2} \ge \frac{a^2}{\tilde{h}\sigma^2} \,\Big|\, (\boldsymbol{x}_1,\ldots,\boldsymbol{x}_n)\right) \\
&= 1 - F_n\left(\frac{a^2}{\tilde{h}\sigma^2}\right)
\end{aligned}
$$

where F_n is the χ^2 distribution function with n degrees of freedom.

Because of (4.23), it holds (see Csörgő and Révész (1981) or Parzen (1975))

$$\sup_{x\in\mathbb{R}}\left|\frac{f_n'(x)}{f_n^2(x)}F_n(x)(1-F_n(x))\right| \le 1$$

hence, because $\tilde{h}_n \le 1$,

$$1 - F_n\left(\frac{a^2}{\tilde{h}\sigma^2}\right) \le 1 - F_n\left(\frac{a^2}{\sigma^2}\right) \tag{4.24}$$

$$\le \frac{(a^2/\sigma^2)^{\frac{n}{2}-1}}{e^{a^2/2\sigma^2}\,2^{n/2}\Gamma(n/2)\left|\frac{1}{2}-\left(\frac{n}{2}-1\right)\frac{\sigma^2}{a^2}\right|}$$

Inserting $a_n = \sigma\sqrt{2\log n}$ in (4.24), we obtain

$$\frac{-\log(1-F_n(2\log n))}{\log n} \ge \frac{n}{2}\left\{1 - \frac{1+\log 2}{\log n} - \frac{\log\log n}{\log n}\right\}$$

$\square$

The basic function of R for the linear regression (corresponding to the least squares method) is `lm`. It is used to fit linear models but it can be used to carry out regression (for example, the analysis of covariance). The call is as follows:

```
> lm(formula, data)
```

where `formula` is a symbolic description of the model (the only required argument) and `data` is an optimal data frame. For example, if we consider the model

$$Y_i = \beta_0 + \beta_1 X_{1i} + \beta_2 X_{2i} + e_i, \quad i = 1,\ldots,n$$

then

```
> lm(y~x1+x2, data)
```

The intercept term is implicitly present; its presence may be confirmed by giving a formula such as y ∼ 1+x1+x2. It may be omitted by giving a −1 term in formula, as in y~x1+x2-1 or also y~0+x1+x2.

lm returns an object of class "lm." Generic functions to perform further operations on this object include, among others: **print** for a simple display, **summary** for a conventional regression analysis output, **coefficients** for extracting the vector of regression coefficients, **residuals** for the residuals (response minus fitted values), **fitted** for fitted mean values, **deviance** for the residual sum of squares, **plot** for diagnostic plots, and **predict** for prediction, including confidence and prediction intervals. **abline** can be used in case of simple linear regression to add the straight regression line to an existing plot.

Consider this example. Koenker and Bassett (1982) give the Engel food expenditure data, which can be found in the package *quantreg*. This is a regression dataset consisting of 235 observations on income (x) and expenditure (y) on food for Belgian working-class households.

```
> library(quantreg)
> data(engel)

> lm(foodexp~income,data=engel)

Call:
lm(formula = foodexp ~ income, data = engel)

Coefficients:
(Intercept)        income
147.4754           0.4852

> engel.fit<-lm(foodexp~income,data=engel)
> plot(foodexp~income,data=engel)
> abline(engel.fit)

> # object of class lm

> ###### print ############
> print(engel.fit)

Call:
lm(formula = foodexp ~ income, data = engel)

Coefficients:
(Intercept)             x
147.4754           0.4852
```

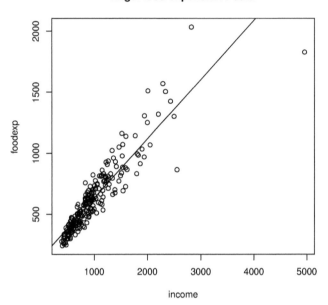

Figure 4.2 *Least square line for Engel food expenditure data.*

```
> ###### summary #########
> summary(engel.fit)

Call:
lm(formula = foodexp ~ income, data = engel)

Residuals:
Min       1Q  Median       3Q      Max
-725.70  -60.24   -4.32    53.41   515.77

Coefficients:
            Estimate Std. Error t value Pr(>|t|)
(Intercept) 147.47539   15.95708   9.242   <2e-16 ***
income        0.48518    0.01437  33.772   <2e-16 ***
---
Signif. codes:  0 '***' 0.001 '**' 0.01 '*' 0.05 '.' 0.1 ' ' 1

Residual standard error: 114.1 on 233 degrees of freedom
Multiple R-squared:  0.8304,Adjusted R-squared:  0.8296
F-statistic:  1141 on 1 and 233 DF,  p-value: < 2.2e-16
```

```
> ###### coefficients ############
> coefficients(engel.fit)
(Intercept)        income
147.4753885     0.4851784

> ###### residuals ###########
> residuals(engel.fit)
1                2                3                4
-95.4873907   -99.1979999   -99.0175287   -54.5459709

5                6                7                8
-16.2232564    27.4411921    80.8752198    77.8958642

.......

229               230              231              232
-30.6916798   -43.5993020   -54.6858594  -110.8549133

233               234              235
38.4621331    14.6014720    89.6828553

> ######### residual sum of squares ##############
> deviance(engel.fit)
[1] 3033805

> ######## fitted values ##############
> fitted(engel.fit)

1          2          3          4          5
351.3268   410.1567   584.6975   457.5433   511.7840

6          7          8          9          10
606.3566   549.8813   622.5450   783.0004   871.5551

.........

226        227        228        229         230
921.4131   524.2629   744.6929   1024.6547   349.0383

231        232        233        234         235
361.2049   410.0542   429.5387   508.0004    660.6373

> ####### diagnostic plots ######
> plot(engel.fit)
```

```
Hit <Return> to see next plot:
Hit <Return> to see next plot:
Hit <Return> to see next plot:
Hit <Return> to see next plot:
```

We obtain Figures 4.3–4.6.

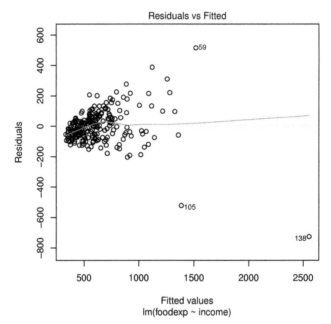

Figure 4.3 *Residuals versus the fitted values for Engel food expenditure data.*

It is clear from the diagnostic plots 4.3 to 4.6 that the dataset contains outliers, so we should look for an alternative more robust method.

4.3 *M*-estimators

The *M*-estimator of parameter $\boldsymbol{\beta}$ in model (4.1) is defined as solutions $\mathbf{M}_n$ of the minimization

$$\sum_{i=1}^{n} \rho(Y_i - \boldsymbol{x}_i'\mathbf{t}) := \min \tag{4.25}$$

with respect to $\mathbf{t} \in \mathbb{R}_p$, where $\rho : \mathbb{R}_1 \mapsto \mathbb{R}_1$ is an absolutely continuous, usually convex function with derivative ψ. Such $\mathbf{M}_n$ is obviously *regression equivariant*, i.e.,

$$\mathbf{M}_n(\boldsymbol{Y} + \boldsymbol{X}\boldsymbol{b}) = \mathbf{M}_n(\boldsymbol{Y}) + \boldsymbol{b} \ \forall \boldsymbol{b} \in \mathbb{R}_p \tag{4.26}$$

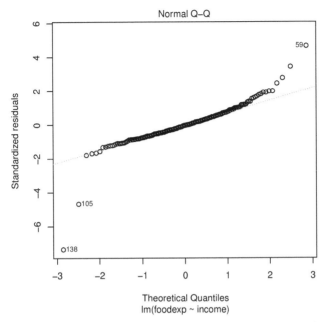

Figure 4.4 *Normal Q-Q plot for Engel food expenditure data.*

but $\mathbf{M}_n$ is generally not *scale equivariant*: generally, it does not hold that

$$\mathbf{M}_n(c\boldsymbol{Y}) = c\mathbf{M}_n(\boldsymbol{Y}) \ \text{ for } \ c > 0 \tag{4.27}$$

A scale equivariant M-estimator we obtain either by a *studentization* or if we estimate the scale simultaneously with the regression parameter. The studentized M-estimator is a solution of the minimization

$$\sum_{i=1}^{n} \rho\left(\frac{Y_i - \boldsymbol{x}_i' \mathbf{t}}{S_n}\right) := \min \tag{4.28}$$

where $S_n = S_n(\boldsymbol{Y}) \geq 0$ is an appropriate scale statistic. To obtain $\mathbf{M}_n$ both regression and scale equivariant, our scale statistic S_n must be scale equivariant and invariant to the regression, i.e.,

$$S_n(c(\boldsymbol{Y} + \boldsymbol{X}\boldsymbol{b})) = cS_n(\boldsymbol{Y}) \ \forall \boldsymbol{b} \in \mathbb{R}_p \ \text{ and } \ c > 0 \tag{4.29}$$

Such is, e.g., the root of the residual sum of squares,

$$S_n(\boldsymbol{Y}) = [(\widehat{\boldsymbol{Y}} - \boldsymbol{Y})'(\widehat{\boldsymbol{Y}} - \boldsymbol{Y})]^{\frac{1}{2}} = [\boldsymbol{Y}'(\boldsymbol{I}_n - \widehat{\boldsymbol{H}})\boldsymbol{Y}]^{\frac{1}{2}}$$

but this is closely connected with the least squares estimator, and thus highly non-robust. Robust scale statistics can be based on the regression quantiles or on the regression rank scores, which will be considered later.

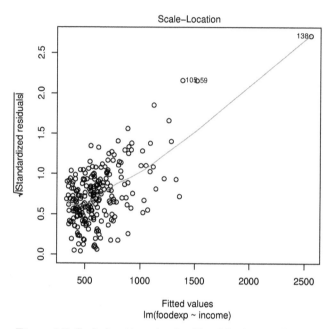

Figure 4.5 *Scale-location plot for Engel food expenditure data.*

The minimization (4.28) should be supplemented with a rule on how to define $\mathbf{M}_n$ in case $S_n(\mathbf{Y}) = 0$; but this mostly happens with probability 0. Moreover, a specific form of the rule does not affect the asymptotic behavior of $\mathbf{M}_n$.

If $\psi(x) = \frac{d\rho(x)}{dx}$ is continuous, then $\mathbf{M}_n$ is a root of the system of equations

$$\sum_{i=1}^{n} \boldsymbol{x}_i \psi \left(\frac{Y_i - \boldsymbol{x}_i' \mathbf{t}}{S_n} \right) = \mathbf{0} \tag{4.30}$$

This system can have more roots, while only one leads to the global minimum of (4.28). Under general conditions, there always exists at least one root of the system (4.30) which is an $\sqrt{n}$-consistent estimator of $\boldsymbol{\beta}$ (see Jurečková and Sen (1996)).

Another important case is that ψ is a nondecreasing step function, hence ρ is a convex, piecewise linear function. Then $\mathbf{M}_n$ is a point of minima of the convex function $\sum_{i=1}^{n} \rho\left((Y_i - \boldsymbol{x}_i' \mathbf{t})/S_n\right)$ over $\mathbf{t} \in \mathbb{R}_p$. In this case, too, we can prove its consistency and asymptotic normality.

If we want to estimate the scale simultaneously with the regression parameter, we can proceed in various ways. One possibility is to consider $(\mathbf{M}_n, \hat{\sigma})$

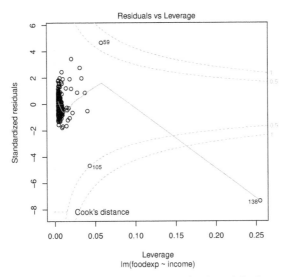

Figure 4.6 *Plot of residuals against leverages for Engel food expenditure data.*

as a solution of the minimization

$$\sum_{i=1}^{n} \sigma\rho\left(\sigma^{-1}(Y_i - x_i't)\right) + a\sigma := \min, \ \mathbf{t} \in \mathbb{R}_p, \ \sigma > 0 \qquad (4.31)$$

with an appropriate constant $a > 0$. We arrive at the system of $p+1$ equations

$$\sum_{i=1}^{n} x_i \psi\left(\frac{Y_i - x_i't}{\sigma}\right) = 0$$

$$\sum_{i=1}^{n} \chi\left(\frac{Y_i - x_i't}{\sigma}\right) = a \qquad (4.32)$$

where $\quad \chi(x) = x\psi(x) - \rho(x) \ \text{ and } \ a = \displaystyle\int_{\mathbb{R}} \chi(x)d\Phi(x)$

and Φ is the standard normal distribution function. The usual choice of ψ is the Huber function (3.16).

The matrix X can be random, fixed or a mixture of random and fixed elements. If the matrix X is random, then its rows are usually independent random vectors, identically distributed, hence they are an independent sample from some multivariate distribution. The influence function of M_n in this situation depends on two arguments, on x and y. Similarly, the possible breakdown and the value of the breakdown point of M_n should be considered not only with respect to changes in observations y, but also with respect to those of x.

4.3.1   Influence function of M-estimator with random matrix

Consider the model (4.1) with random matrix $\boldsymbol{X}$, where $(\boldsymbol{x}'_i, Y_i)'$, $i = 1, \ldots, n$ are independent random vectors with values in $\mathbb{R}_p \times \mathbb{R}_1$, identically distributed with distribution function $P(\boldsymbol{x}, y)$. If ρ has an absolutely continuous derivative ψ, then the statistical functional $\mathbf{T}(P)$, corresponding to the estimator (4.25), is a solution of the system of p equations

$$\int_{\mathbb{R}_{p+1}} \boldsymbol{x}\psi(y - \boldsymbol{x}'\mathbf{T}(P))dP(\mathbf{x}, y) = \mathbf{0} \qquad (4.33)$$

Let P_t denote the contaminated distribution

$$P_t = (1 - t)P + t\delta(\boldsymbol{x}_0, y_0), \ 0 \le t \le 1, \ (\mathbf{x}_0, y_0) \in \mathbb{R}_p \times \mathbb{R}$$

where $\delta(\boldsymbol{x}_0, y_0)$ is a degenerated distribution with the probability mass concentrated in the point $(\boldsymbol{x}_0, y_0)$. Then the functional $\mathbf{T}(P_t)$ solves the system of equations

$$(1 - t) \int_{\mathbb{R}_{p+1}} \boldsymbol{x}\psi(y - \boldsymbol{x}'\mathbf{T}(P_t))dP(\boldsymbol{x}, y)$$

$$+ t\boldsymbol{x}_0\psi(y_0 - \boldsymbol{x}'_0\mathbf{T}(P_t)) = \mathbf{0}.$$

Differentiating by t we obtain

$$-\int_{\mathbb{R}_{p+1}} \boldsymbol{x}\psi(y - \boldsymbol{x}'\mathbf{T}(P_t))dP(\boldsymbol{x}, y) + \boldsymbol{x}_0\psi(y_0 - \boldsymbol{x}'_0\mathbf{T}(P_t))$$

$$-(1 - t) \int_{\mathbb{R}_{p+1}} \boldsymbol{x}'\boldsymbol{x}\frac{d\mathbf{T}(P_t)}{dt}\psi'(y - \boldsymbol{x}'\mathbf{T}(P_t))dP(\mathbf{x}, y)$$

$$-t\boldsymbol{x}'_0\boldsymbol{x}_0\frac{d\mathbf{T}(P_t)}{dt}\psi'(y_0 - \boldsymbol{x}'_0\mathbf{T}(P_t)) = \mathbf{0}$$

and we get the influence function

$$\mathbf{IF}(\boldsymbol{x}_0, y_0; \mathbf{T}, P) = \frac{d\mathbf{T}(P_t)}{dt}\bigg|_{t=0}$$

if we put $t = 0$ and notice that, on account of (4.33),

$$\int_{\mathbb{R}_{p+1}} \boldsymbol{x}\psi(y - \boldsymbol{x}'\mathbf{T}(P_t))dP(\boldsymbol{x}, y) = \mathbf{0}$$

then

$$\mathbf{IF}(\boldsymbol{x}_0, y_0; \mathbf{T}, P) \int_{\mathbb{R}_{p+1}} \boldsymbol{x}'\boldsymbol{x}\psi'(y - \boldsymbol{x}'\mathbf{T}(P))dP(\boldsymbol{x}, y)$$

$$= \boldsymbol{x}_0\psi(y_0 - \boldsymbol{x}'_0\mathbf{T}(P))$$

Hence, the influence function of the *M*-estimator is of the form

$$\mathbf{IF}(\boldsymbol{x}_0, y_0; \mathbf{T}, P) = \mathbf{B}^{-1}\boldsymbol{x}_0\psi(y_0 - \boldsymbol{x}_0'\mathbf{T}(P)) \qquad (4.34)$$

where

$$\mathbf{B} = \int_{\mathbb{R}^{p+1}} \boldsymbol{x}'\boldsymbol{x}\psi'(y - \boldsymbol{x}'\mathbf{T}(P))dP(\boldsymbol{x}, y) \qquad (4.35)$$

Observe that the influence function (4.35) is bounded in y_0, provided ψ is bounded; however, it is unbounded in $\boldsymbol{x}_0$, and thus the *M*-estimator is non-robust with respect to $\boldsymbol{X}$. Many authors tried to overcome this shortage and introduced various generalized *M*-estimators, called *GM*-estimators, that out-perform the effect of outliers in $\boldsymbol{x}$ with the aid of properly chosen weights.

4.3.2 Large sample distribution of the M-estimator with nonrandom matrix

Because the *M*-estimator is nonlinear and defined implicitly as a solution of a minimization, it would be very difficult to derive its distribution function under a finite number of observations. Moreover, even if we were able to derive this distribution function, its complicated form would not give the right picture of the estimator. This applies also to other robust estimators. In this situation, we take recourse in the limiting (asymptotic) distributions of estimators, which are typically normal, and their covariance matrices fortunately often have a compact form. Deriving the asymptotic distribution is not easy and we should use various, sometimes nontraditional methods that have an interest of their own, but their details go beyond this text. In this context, we refer to the monographs cited in the literature, such as Huber (1981), Hampel et al. (1986), Rieder (1994), Jurečková and Sen (1996), among others.

Let us start with the asymptotic properties of the non-studentized estima-tor with a nonrandom matrix $\boldsymbol{X}$. Assume that the distribution function F of errors U_i in model (4.1) is symmetric around zero. Consider the *M*-estimator $\mathbf{M}_n$ as a solution of the minimization (4.25) with an odd, absolutely continuous function $\psi = \rho'$ such that $\mathbb{E}_F\psi^2(U_1) < \infty$. The matrix $\boldsymbol{X} = \boldsymbol{X}_n$ is supposed to be of rank p and $\max_{1 \le i \le n} h_{ii}^{(n)} \to 0$ as $n \to \infty$, where $h_{ii}^{(n)}$ is the maximal diagonal element of the projection (hat) matrix $\widehat{\boldsymbol{H}}_n = \boldsymbol{X}_n(\boldsymbol{X}_n'\boldsymbol{X}_n)^{-1}\boldsymbol{X}_n'$, then

$$\mathbf{M}_n \xrightarrow{p} \boldsymbol{\beta} \qquad (4.36)$$

$$\mathcal{L}\left\{(\boldsymbol{X}_n'\boldsymbol{X}_n)^{\frac{1}{2}}(\mathbf{M}_n - \boldsymbol{\beta})\right\} \to \mathcal{N}_p\left(\mathbf{0}, \sigma^2(\psi, F)\mathbf{I}_p\right)$$

$$\text{as } n \to \infty, \text{ where } \sigma^2(\psi, F) = \frac{\mathbb{E}_F\psi^2(U_1)}{(\mathbb{E}_F\psi'(U_1))^2}$$

If, moreover, $\frac{1}{n}\boldsymbol{X}_n'\boldsymbol{X}_n \to \boldsymbol{Q}$, where $\boldsymbol{Q}$ is a positively definite matrix of order $p \times p$, then

$$\mathcal{L}\left\{\sqrt{n}(\mathbf{M}_n - \boldsymbol{\beta})\right\} \to \mathcal{N}_p\left(\mathbf{0}, \sigma^2(\psi, F)\boldsymbol{Q}^{-1}\right)$$

If ψ has jump points, but is nondecreasing, and F is absolutely continuous with density f, then (4.36) is still true with the only difference being that

$$\sigma^2(\psi, F) = \frac{\mathbb{E}_F \psi^2(U_1)}{(\int_\mathbb{R} f(x) d\psi(x))^2}$$

Notice that $\sigma^2(\psi, F)$ is the same as in the asymptotic distribution of the M-estimator of location.

If M-estimator $\mathbf{M}_n$ is studentized by the scale statistic S_n such that $S_n \xrightarrow{p} S(F)$ as $n \to \infty$, then the asymptotic covariance matrix of $\mathbf{M}_n$ depends on $S(F)$. In the models with intercepts, only the first component of the estimator is asymptotically affected by $S(F)$.

4.3.3 Large sample properties of the M-estimator with random matrix

If the system of equations

$$\mathbb{E}_P[\boldsymbol{x}\psi(y - \boldsymbol{x}'\mathbf{t}) = \mathbf{0}$$

has a unique root $\mathbf{T}(P) = \boldsymbol{\beta}$, then

$$\mathbf{T}(P_n) \to \mathbf{T}(P)$$

as $n \to \infty$, where P_n is the empirical distribution pertaining to observations $((\boldsymbol{x}_1, y_1), \ldots, (\mathbf{x}_n, y_n))$. The functional $\mathbf{T}(P_n)$ admits, under some conditions on probability distribution P, the following asymptotic representation

$$\mathbf{T}(P_n) = \mathbf{T}(P) + \frac{1}{n}\mathbf{IF}(\boldsymbol{x}, y; \mathbf{T}, P) + \mathbf{o}_p(n^{-\frac{1}{2}}) \quad \text{as } n \to \infty$$

If $\mathbb{E}_P \|\mathbf{IF}(\boldsymbol{x}, y; \mathbf{T}, P)\|^2 < \infty$, the above representation further leads to the asymptotic distribution of $\mathbf{T}(P_n)$:

$$\mathcal{L}\left\{\sqrt{n}(\mathbf{T}(P_n) - \mathbf{T}(P))\right\} \to \mathcal{N}_p(\mathbf{0}, \boldsymbol{\Sigma}) \tag{4.37}$$

where

$$\boldsymbol{\Sigma} = \mathbb{E}_P[\mathbf{IF}(\boldsymbol{x}, y; \mathbf{T}, P)]'[\mathbf{IF}(\boldsymbol{x}, y; \mathbf{T}, P)] = \mathbf{B}^{-1}\mathbf{A}\mathbf{B}^{-1}$$

$\mathbf{B}$ is the matrix defined in (4.35) and

$$\mathbf{A} = \int_{\mathbb{R}_{p+1}} \boldsymbol{x}'\boldsymbol{x}\psi^2(y - \mathbf{x}'\mathbf{T}(P))dP(\boldsymbol{x}, y)$$

The package MASS implements the procedure rlm(), which corresponds to either the M-estimator or MM-estimator. The package MASS is recommended; it will be distributed with every binary distribution of R. The arguments of the function rlm() were described in a previous chapter. The Huber, bisquare and Hampel M-estimators are calculated, see Figure 4.7. Alternatively we can also use the function lmrob in the package robustbase.

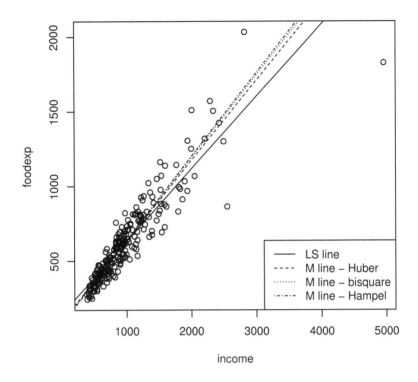

Figure 4.7 *Plot of M-estimate lines for Engel food expenditure data.*

```
> library(MASS)
> library(quantreg)
> data(engel)
> rlm(foodexp~income,data=engel)
Call:
rlm(formula = foodexp ~ income, data = engel)
Converged in 10 iterations

Coefficients:
(Intercept)      income
99.4319013    0.5368326

Degrees of freedom: 235 total; 233 residual
Scale estimate: 81.4
> summary(rlm(foodexp~income,data=engel))
```

```
Call: rlm(formula = foodexp ~ income, data = engel)
Residuals:
Min        1Q    Median       3Q       Max
-933.748  -54.995    4.768   53.714   418.020


Coefficients:
Value    Std. Error t value
(Intercept) 99.4319 12.1244       8.2010
income        0.5368  0.0109      49.1797


Residual standard error: 81.45 on 233 degrees of freedom
> deviance(rlm(foodexp~income,data=engel))
[1] 3203849


> rlm(foodexp~income,data=engel, psi=psi.bisquare )
Call:
rlm(formula = foodexp ~ income, data = engel,
psi = psi.bisquare)
Converged in 11 iterations


Coefficients:
(Intercept)        income
86.245026     0.552921


Degrees of freedom: 235 total; 233 residual
Scale estimate: 81.3


> rlm(foodexp~income,data=engel, psi=psi.hampel, init="ls")
Call:
rlm(formula = foodexp ~ income, data = engel, psi = psi.hampel,
init = "ls")
Converged in 8 iterations


Coefficients:
(Intercept)        income
80.2517611    0.5601229


Degrees of freedom: 235 total; 233 residual
Scale estimate: 80.4

plot(foodexp~income,data=engel)
abline(init)
abline(rlm(foodexp~income,data=engel),lty=2)
abline(rlm(foodexp~income,data=engel, psi=psi.bisquare ), lty=3)
abline(rlm(foodexp~income,data=engel, psi=psi.hampel ), lty=4)
```

```
legend("bottomright",c("LS line", "M line - Huber",
  "M line - bisquare", "M line - Hampel" ), lty = c(1,2,3,4))

> library(robustbase)
> init<-lm(foodexp~income,data=engel)
> lmrob(foodexp~income, data=engel, control = lmrob.control(
+          method = "M",psi = "hampel", tuning.psi = c(2,4,8)),
+          init = list(coefficients = init$coef,
+                         scale = mad(resid(init))) )

Call:
lmrob(formula = foodexp ~ income, data = engel, control =
lmrob.control(method = "M", psi = "hampel", tuning.psi =
c(2, 4, 8)), init = list(coefficients = init$coef,
scale = mad(resid(init))))

Coefficients:
(Intercept)        income
80.7900            0.5594
```

4.4   *GM*-estimators

The influence function (4.34) of the M-estimator is unbounded in x, thus the M-estimator is sensitive to eventual leverage points in matrix X. The choice of function ψ has no effect on this phenomenon. Some authors proposed to supplement the definition of the M-estimator by suitable weights w that reduce the influence of the gross values of x_{ij}.

Mallows (1973, 1975) proposed the generalized M-estimator as a solution of the minimization

$$\sum_{i=1}^{n} \sigma w(\boldsymbol{x}_i)\rho\left(\frac{Y_i - \mathbf{x}_i'\mathbf{t}}{\sigma}\right) := \min, \ \mathbf{t} \in \mathbb{R}_p, \sigma > 0 \tag{4.38}$$

If $\psi = \rho'$ is continuous, then the generalized M-estimator solves the equation

$$\sum_{i=1}^{n} \boldsymbol{x}_i w(\boldsymbol{x}_i)\psi\left(\frac{Y_i - \mathbf{x}_i'\mathbf{t}}{\sigma}\right) = \mathbf{0} \tag{4.39}$$

The influence function of the pertaining functional $\mathbf{T}(P)$ then equals

$$\mathbf{IF}(\boldsymbol{x}, y; \mathbf{T}, P) = \mathbf{B}^{-1}\boldsymbol{x}w(\boldsymbol{x})\psi\left(\frac{y - \boldsymbol{x}'\mathbf{T}(P)}{S(P)}\right) \tag{4.40}$$

where $S(P)$ is the functional corresponding to the solution σ in the minimization (4.38). We obtain a bounded influence function if we take w leading to bounded $\boldsymbol{x}w(\boldsymbol{x})$.

Such an estimator is a special case of the following GM-estimator, which solves the equation

$$\sum_{i=1}^{n} \eta\left(\boldsymbol{x}_i, \frac{Y_i - \mathbf{x}_i' \mathbf{t}}{\sigma}\right) = \mathbf{0}$$

(4.41)

$$\sum_{i=1}^{n} \chi\left(\frac{Y_i - \mathbf{x}_i' \mathbf{t}}{\sigma}\right) = \mathbf{0}$$

with functions η, χ, where $\eta : \mathbb{R}_p \times \mathbb{R} \mapsto \mathbb{R}$ and $\chi : \mathbb{R} \mapsto \mathbb{R}$.

If we take

$$\eta(\boldsymbol{x}, u) = u \quad \text{and} \quad \chi(u) = u^2 - 1$$

we obtain the least squares estimator: the choice $\eta(\boldsymbol{x}, u) = \psi(u)$ leads to the M-estimator, and

$$\eta(\boldsymbol{x}, u) = w(\boldsymbol{x})\psi(u)$$

leads to the Mallows GM-estimator. The usual choice of the function η is $\eta(\boldsymbol{x}, u) = \frac{\psi_1(\mathbf{x})}{\|\boldsymbol{x}\|}\psi(u)$, where ψ is, e.g., the Huber function. The choice of function χ usually coincides with (4.32).

The statistical functionals $\mathbf{T}(P)$ and $S(P)$ corresponding to $\mathbf{M}_n$ and σ_n are defined implicitly as a solution of the system of equations:

$$\int_{\mathbb{R}_{p+1}} \boldsymbol{x} \, \eta\left(\boldsymbol{x}, \frac{y - \mathbf{x}' \mathbf{T}(P)}{S(P)}\right) dP(\boldsymbol{x}, y) = \mathbf{0}$$

(4.42)

$$\int_{\mathbb{R}_{p+1}} \chi\left(\frac{y - \mathbf{x}' \mathbf{T}(P)}{S(P)}\right) dP(\boldsymbol{x}, y) = \mathbf{0}$$

The influence function of functional $\mathbf{T}(P)$ in the special case $\sigma = 1$ has the form

$$\mathbf{IF}(\boldsymbol{x}, y; \mathbf{T}, P) = \mathbf{B}^{-1}\boldsymbol{x}\eta(\boldsymbol{x}, y - \mathbf{x}'\mathbf{T}(P)),$$

where

$$\mathbf{B} = \int_{\mathbb{R}_{p+1}} \boldsymbol{x}'\boldsymbol{x} \left[\frac{\partial}{\partial u}\eta(\boldsymbol{x}, u)\right]_{u = y - \mathbf{x}'\mathbf{T}(P)} dP(\boldsymbol{x}, y)$$

The asymptotic properties of GM-estimators were studied by Maronna and Yohai (1981), among others. Under some regularity conditions, the GM-estimators are strongly consistent and $\sqrt{n}(\mathbf{T}(P_n) - \mathbf{T}(P))$ has asymptotic p-dimensional normal distribution $\mathcal{N}_p(\mathbf{0}, \boldsymbol{\Sigma})$ with covariance matrix $\boldsymbol{\Sigma} = \mathbf{B}^{-1}\mathbf{A}\mathbf{B}^{-1}$, where

$$\mathbf{A} = \int_{\mathbb{R}_{p+1}} \boldsymbol{x}'\boldsymbol{x}\eta^2(\mathbf{x}, y - \boldsymbol{x}'\mathbf{T}(P))dP(\boldsymbol{x}, y)$$

Krasker and Welsch (1982) proposed a *GM*-estimator as a solution of the system of equations

$$\sum_{i=1}^{n} \boldsymbol{x}_i w_i \frac{Y_i - \boldsymbol{x}_i' \mathbf{t}}{\sigma} = \mathbf{0}$$

with weights $w_i = w(\boldsymbol{x}_i, Y_i, \mathbf{t}) > 0$ determined so that they maximize the asymptotic efficiency of the estimator (with respect to the asymptotic covariance matrix $\boldsymbol{\Sigma}$) under the constraint $\gamma^* \leq a < \infty$, where γ^* is the global sensitivity of the functional $\mathbf{T}$ under distribution P, i.e.,

$$\gamma^* = \sup_{\boldsymbol{x}, y} \left[(\mathbf{IF}(\boldsymbol{x}, y; \mathbf{T}, P))' \, \boldsymbol{\Sigma}^{-1} \, (\mathbf{IF}(\boldsymbol{x}, y; \mathbf{T}, P)) \right]^{\frac{1}{2}}$$

As a solution, we obtain the weights of the form

$$w(\boldsymbol{x}, y, \mathbf{t}) = \min \left\{ 1, \frac{a}{\left| \frac{y - \mathbf{x}'\mathbf{t}}{\sigma} \right| (\boldsymbol{x}' \mathbf{A} \mathbf{x})^{\frac{1}{2}}} \right\}$$

where

$$\mathbf{A} = \int_{\mathbb{R}_{p+1}} \boldsymbol{x}' \boldsymbol{x} \left(\frac{y - \boldsymbol{x}' \mathbf{t}}{\sigma} \right) w^2(\boldsymbol{x}, y, \mathbf{t}) dP(\boldsymbol{x}, y)$$

The Krasker-Welsch estimator has a bounded influence function, but it should be computed iteratively, because matrix $\mathbf{A}$ depends on w.

For computation of the *GM*-estimator (4.38) we can again use the procedure `rlm()` from the package MASS in the following way. For the example of food expenditure data the weights can be set, e.g., to $w_i = 1/income_i^2$ or $w_i = 1 - h_{ii}$, where h_{ii} is the leverage score, the i-th diagonal element of the project (*hat*) matrix $\widehat{\boldsymbol{H}} = \boldsymbol{X} \left(\boldsymbol{X}' \boldsymbol{X} \right)^{-1} \boldsymbol{X}'$.

```
> library(MASS)
> library(quantreg)
> data(engel)
> rlm(foodexp~income, data=engel, weights=1/income^2,
+     wt.method="case")
Call:
rlm(formula = foodexp ~ income, data = engel, weights
= 1/income^2, wt.method = "case")
Converged in 9 iterations

Coefficients:
(Intercept)      income
48.5745393    0.6025803

Degrees of freedom: 235 total; 233 residual
Scale estimate: 54.6
> xhat <- 1-hat(model.matrix(foodexp~income, data=engel))
```

```
> rlm(foodexp~income, data=engel, weights=xhat,
+      wt.method="case")
Call:
rlm(formula = foodexp ~ income, data = engel, weights = xhat,
wt.method = "case")
Converged in 10 iterations


Coefficients:
(Intercept)        income
95.7240778    0.5413302


Degrees of freedom: 235 total; 233 residual
Scale estimate: 81.6
```

4.5 R-estimators, GR-estimators

Consider the linear model with an intercept

$$Y_i = \theta + \mathbf{x}_i^\top \boldsymbol{\beta} + e_i, \quad i = 1, \ldots, n \tag{4.43}$$

where $\mathbf{Y} = (Y_1, \ldots, Y_n)^\top$ are observed responses, $\mathbf{x}_i = (x_{i1}, \ldots, x_{ip})^\top$, $1 \le i \le n$ are observable regressors, $\boldsymbol{\beta} = (\beta_1, \ldots, \beta_p)^\top$ is a vector of regression (slopes) parameters, and θ is an intercept. We assume that the e_i are i.i.d. errors with a continuous (but unknown) distribution function F. When estimating the parameters with the aid of ranks, we should distinguish between estimating the slopes $\boldsymbol{\beta}$ and estimating the intercept θ. Estimating $\boldsymbol{\beta}$ is more straightforward and is based on ranks of residuals; estimating θ is based on a signed rank statistics, but supplemented with an estimate of the nuisance $\boldsymbol{\beta}$ (aligned signed-rank statistics).

We shall concentrate mainly on estimating of the slopes $\boldsymbol{\beta}$. For every $\mathbf{b} \in \mathbf{R}_p$, let

$$R_{ni}(\mathbf{b}) = \text{Rank of } (Y_i - \mathbf{x}_i^\top \mathbf{b}) \text{ among the } Y_j - \mathbf{x}_j^\top \mathbf{b} \ (1 \le j \le n), \tag{4.44}$$

for $i = 1, \ldots, n$. The ranks $R_{ni}(\mathbf{b})$ are translation invariant, and hence independent of unknown θ. Let $a_n(1) \le \ldots \le a_n(n)$ be a set of scores (not all equal), as in the location model. The estimate of $\boldsymbol{\beta}$ is based on a vector of linear rank statistics

$$\mathbf{S}_n(\mathbf{b}) = (S_{n1}(\mathbf{b}), \ldots, S_{np}(\mathbf{b}))^\top$$
$$= \sum_{i=1}^n (\mathbf{x}_i - \bar{\mathbf{x}}_n) a_n(R_{ni}(\mathbf{b})), \quad \mathbf{b} \in \mathbf{R}_p, \tag{4.45}$$

where $\bar{\mathbf{x}}_n = n^{-1} \sum_{i=1}^n \mathbf{x}_i$. If $\boldsymbol{\beta} = \mathbf{0}$, then $R_{n1}(\mathbf{0}), \ldots R_{nn}(\mathbf{0})$ are interchangeable random variables [assuming each permutation of $1, \ldots, n$ with the common probability $(n!)^{-1}$], and we have

$$E_{\beta=0}(\mathbf{S}_n(\mathbf{0})) = \mathbf{0}, \quad \forall n \ge 1.$$

This leads to an idea to "equate" $\mathbf{S}_n(\mathbf{b})$ to $\mathbf{0}$, to obtain a suitable estimator of $\boldsymbol{\beta}$. However, the statistics are step functions, hence this system of equations may not have a solution.

Instead of that, we may define the R-estimator of $\boldsymbol{\beta}$ by the minimization of a suitable norm of $\mathbf{S}_n(\mathbf{b})$, for instance of the L_1-norm

$$\|\mathbf{S}_n(\mathbf{b})\| = \sum_{j=1}^{p} |S_{nj}(\mathbf{b})|, \ \mathbf{b} \in \boldsymbol{R}_p$$

and define

$$\widetilde{\mathbf{b}} : \| \mathbf{S}_n(\widetilde{\mathbf{b}})\| = \inf_{\mathbf{b} \in R_p} \|\mathbf{S}_n(\mathbf{b})\|. \tag{4.46}$$

Such $\widetilde{\mathbf{b}}$ may not be uniquely determined, and we can have the set

$$\mathcal{D}_n = \text{ set of all } \widetilde{\mathbf{b}} \text{ satisfying (4.46)}. \tag{4.47}$$

We can eliminate the arbitrariness in (4.46) by letting $\widehat{\boldsymbol{\beta}}_n$ being the center of gravity of $\mathcal{D}_n$.

A slightly different formulation was proposed by Jaeckel (1972) who used the following measure of *rank dispersion*

$$D_n(\mathbf{b}) = \sum_{i=1}^{n} (Y_i - \mathbf{x}_i^\top \mathbf{b}) a_n(R_{ni}(\mathbf{b})), \ \ \mathbf{b} \in \boldsymbol{R}_p \tag{4.48}$$

where $a_n(1), \ldots, a_n(n)$ are again nondecreasing scores. Jaeckel showed that $D_n(\mathbf{b})$ is nonnegative, continuous, piecewise linear and convex function of $\mathbf{b} \in \boldsymbol{R}_p$

and proposed to estimate $\boldsymbol{\beta}$ by minimizing $D_n(\mathbf{b})$ with respect to $\mathbf{b} \in \boldsymbol{R}_p$. If we set $\bar{a}_n = n^{-1} \sum_{i=1}^{n} a_n(i) = 0$, without loss of generality, we can show that $D_n(\mathbf{b})$ is translation-invariant. Indeed, if we denote $D_n(\mathbf{b}, k)$ as the measure (4.48) calculated for the pseudo-observations $Y_i + k - \mathbf{b}^\top \mathbf{x}_i$, $i = 1, \ldots, n$, we can write

$$\begin{aligned} D_n(\mathbf{b}, k) &= \sum_{i=1}^{n} (Y_i + k - \mathbf{b}^\top \mathbf{x}_i)(a_n(R_{ni}(\mathbf{b})) - \bar{a}_n) \\ &= D_n(\mathbf{b}) + k \sum_{i=1}^{n} (a_n(R_{ni}(\mathbf{b})) - \bar{a}_n) = D_n(\mathbf{b}) \end{aligned}$$

because $a_n(R_{ni}(\mathbf{b}))$ are translation-invariant. Hence $D_n(\mathbf{b})$ is differentiable in $\mathbf{b}$ almost everywhere and

$$(\partial/\partial \mathbf{b}) D_n(\mathbf{b}) \Big|_{\mathbf{b}^0} = -\mathbf{S}_n(\mathbf{b}^0) \tag{4.49}$$

at any point $\mathbf{b}^0$ of differentiability of D_n. If D_n is not differentiable in $\mathbf{b}^0$, we can work with the subgradient $\nabla D_n(\mathbf{b}_0)$ of D_n at $\mathbf{b}_0$ defined as the operation satisfying

$$D_n(\mathbf{b}) - D_n(\mathbf{b}^0) \geq (\mathbf{b} - \mathbf{b}^0)\nabla D_n(\mathbf{b}_0) \qquad (4.50)$$

for all $\mathbf{b} \in \mathbf{R}_p$.

The R-estimator $\widehat{\boldsymbol{\beta}}_n$ is defined implicitly as a solution of a computationally demanding problem, which should be solved iteratively. However asymptotically, with increasing number n of observations, we can see its advantages: The R-estimator, as an estimator of the slopes, is invariant to the shift in location and regression, and scale equivariant. If F is known, we can determine the score function $\varphi(F)$ leading to an asymptotically efficient estimator of the slope components, which is asymptotically equivalent to the maximum likelihood one. Moreover, with $\varphi(t) = \psi(F^{-1}(t))$, $0 < t < 1$, the R-estimator with score-function φ is asymptotically equivalent to M-estimator of $\boldsymbol{\beta}$, generated by score-function ψ. Various other relations are apparent from the following asymptotic approximation (representation) of $\widehat{\boldsymbol{\beta}}_n$ with the aid of the sum of independent summands (the detailed proof is, e.g., in Jurečková, Sen and Picek (2013)):

Proposition 4.1 *Assume that*

- *the distribution of the errors e_i has an absolutely continuous density f and finite Fisher information;*
- *the function $\varphi : (0,1) \mapsto \mathbf{R}_1$ is non-constant, nondecreasing and square integrable on $(0,1)$, i.e., $0 < \int_0^1 \varphi^2(u)du < \infty$.*
- *Assume that*

$$\lim_{n \to \infty} \max_{1 \leq i \leq n} (\mathbf{x}_i - \bar{\mathbf{x}}_n)^\top \mathbf{V}_n^{-1}(\mathbf{x}_i - \bar{\mathbf{x}}_n) = 0, \qquad (4.51)$$

$$\text{where} \quad \mathbf{V}_n = \sum_{i=1}^n (\mathbf{x}_i - \bar{\mathbf{x}}_n)(\mathbf{x}_i - \bar{\mathbf{x}}_n)^\top, \quad n > p,$$

[Generalized Noether condition].

Then, as $n \to \infty$,

$$\widehat{\boldsymbol{\beta}}_n - \boldsymbol{\beta} = \gamma^{-1}\mathbf{V}_n^{-1}\sum_{i=1}^n (\mathbf{x}_i - \bar{\mathbf{x}}_n)\varphi(F(e_i)) + o_p\left(\|\mathbf{V}_n^{-\frac{1}{2}}\|\right), \qquad (4.52)$$

$$\text{where} \quad \gamma = \int_{-\infty}^{\infty} \varphi(F(x))\{-f'(x)\}dx.$$

The function `rfit` from the package *Rfit* can be used to minimize Jaeckel's dispersion function (4.48). By default the Wilcoxon (linear) scores are used, but user-defined scores can be supplied. As an estimate of the intercept, the median of residuals is provided by default. The function `disp` returns the resulting rank dispersion.

```
> library(quantreg)
> data(engel)
> library(Rfit)
> (r.fit <- rfit(foodexp~income, data=engel))
Call:
rfit.default(formula = foodexp ~ income, data = engel)

Coefficients:
(Intercept)      income
103.6453126   0.5377772

> disp(r.fit$coef[-1], engel$income, engel$foodexp,
+     scores = wscores)
[,1]
[1,] 22708.66
```

4.6   *L*-estimators, regression quantiles

L-estimators of the location parameter as linear combinations of order statistics or linear combinations of functions of order statistics are highly appealing
and intuitive, because they are formulated explicitly, not as solutions of minimization problems or of systems of equations. The calculation of *L*-estimators
is much easier. Naturally, many statisticians tried to extend the *L*-estimators
to the linear regression model. Surprisingly, this extension is not easy, because
it was difficult to find a natural and intuitive extension of the empirical (sample) quantile to the regression model. A successful extension of the sample
quantile appeared only in 1978, when Koenker and Bassett introduced the
regression α-quantile $\widehat{\boldsymbol{\beta}}(\alpha)$ for model (4.1). It is more illustrative in a model
with an intercept: hence let us assume that β_1 is an intercept and that matrix
$\boldsymbol{X}$ satisfies the condition

$$x_{i1} = 1, \ i = 1, \ldots, n \tag{4.53}$$

The regression α-quantile $\widehat{\boldsymbol{\beta}}(\alpha)$, $0 < \alpha < 1$ is defined as a solution of the
minimization

$$\sum_{i=1}^{n} \rho_\alpha(Y_i - \boldsymbol{x}_i'\mathbf{t}) := \min, \ \mathbf{t} \in \mathbb{R}_p \tag{4.54}$$

with the criterion function

$$\rho_\alpha(x) = |x|\{\alpha I[x > 0] + (1 - \alpha)I[x < 0]\}, \ x \in \mathbb{R} \tag{4.55}$$

The function ρ_α is piecewise linear and convex; hence we can intuitively expect that the minimization (4.54) can be solved by some modification of the

simplex algorithm of the linear programming. Indeed, Koenker and Bassett (1978) proposed to calculate $\widehat{\beta}(\alpha)$ as the component β of the optimal solution $(\beta, \mathbf{r}^+, \mathbf{r}^-)$ of the parametric linear programming problem

$$\alpha \sum_{i=1}^{n} r_i^+ + (1 - \alpha) \sum_{i=1}^{n} r_i^- : \min \qquad (4.56)$$

under constraint
$$\sum_{j=1}^{p} x_{ij}\beta_j + r_i^+ - r_i^- = Y_i, \ i = 1, \ldots, n$$

$$\beta_j \in \mathbb{R}_1, \ j = 1, \ldots, p, \ r_i^+, r_i^- \geq 0, \ i = 1, \ldots, n,$$

$$0 < \alpha < 1$$

The variables r_i^+ and r_i^- in (4.56) are equal to the positive and negative parts of residuals $Y_i - \boldsymbol{x}_i'\boldsymbol{\beta}, \ i = 1, \ldots, n$.

The linear programming problem (4.56) not only enables calculating the regression quantiles with the aid of the simplex algorithm, but it also illustrates the structure of the regression quantiles. It is known from linear programming theory that the set $B(\alpha)$ of solutions of (4.56) (and thus also of (4.54)) is nonempty, compact and polyhedral. We can choose $\widehat{\beta}(\alpha)$ as a lexicographically maximal element $B(\alpha)$, unless we have some other rule prescribed. Being considered as a function of α, the regression quantile $\widehat{\beta}(\alpha)$ is a step function of $\alpha \in (0, 1)$.

The population counterpart (i.e., the corresponding statistical functional) of $\widehat{\beta}(\alpha)$ is the *population regression quantile*

$$\boldsymbol{\beta}(\alpha) = (\beta_1 + F^{-1}(\alpha), \beta_2, \ldots, \beta_p)' \qquad (4.57)$$

The asymptotic properties of $\widehat{\beta}(\alpha)$ are analogous to those of sample quantiles in the location model. Indeed, $\sqrt{n}(\widehat{\beta}_n(\alpha) - \beta(\alpha))$ has a p-dimensional asymptotic normal distribution

$$\mathcal{N}_p \left(\mathbf{0}, \frac{\alpha(1 - \alpha)}{(f(F^{-1}(\alpha))^2} \boldsymbol{Q}^{-1} \right) \qquad (4.58)$$

under some conditions on matrix $\boldsymbol{X}_n$ and on distribution function F of errors in model (4.1). For instance, (4.58) holds if $\boldsymbol{X}_n$ is either fixed (nonrandom) with $\lim_{n \to \infty} \frac{1}{n} \boldsymbol{X}_n' \boldsymbol{X}_n = \boldsymbol{Q}$, or random (up to the first column that corresponds to the intercept) and $\lim_{n \to \infty} \mathbb{E} \mathbf{x}_1 \boldsymbol{x}_1' = \boldsymbol{Q}$, where $\boldsymbol{Q}$ is a positively definite matrix of order $p \times p$, and if F is symmetric around 0, strictly increasing and has a positive derivative f in a neighborhood of $F^{-1}(\alpha)$. This is in accord with the asymptotic distribution of the sample α-quantile (whose matrix is $\boldsymbol{X} = \mathbf{1}_n = (1, \ldots, 1)' \in \mathbb{R}_n$).

The regression quantiles provide a basis for various L-estimators of parameter $\boldsymbol{\beta}$ in a linear regression model. The most popular is the L_1-estimator, or regression median, which is the regression α-quantile with $\alpha = \frac{1}{2}$. A broad class of *L*-estimators is linear combinations of a finite number of regression quantiles. From a practical point of view, the *trimmed least-squares estimator* proposed by Koenker and Bassett (1978) is very appealing. This is a straightforward extension of the trimmed mean to the linear regression model and is defined as follows: fix α_1, α_2, $0 < \alpha_1 < \alpha_2 < 1$, $i = 1, \ldots, n$, put

$$a_i = I\left[\boldsymbol{x}_i'\widehat{\boldsymbol{\beta}}_n(\alpha_1) < Y_i < \boldsymbol{x}_i'\widehat{\boldsymbol{\beta}}_n(\alpha_2)\right], \ i = 1, \ldots, n \qquad (4.59)$$

and calculate the weighted least squares estimator with the weights a_i. This estimator $\mathbf{T}_n(\alpha_1, \alpha_2)$, called the (α_1, α_2)-trimmed least squares estimator, can be written in an explicit form

$$\mathbf{T}_n(\alpha_1, \alpha_2) = (\boldsymbol{X}_n'\mathbf{A}_n\mathbf{X}_n)^{-1}\boldsymbol{X}_n'\mathbf{A}_n\boldsymbol{Y}_n \qquad (4.60)$$

where $\mathbf{A}_n = \text{diag}(a_i)$ is a diagonal matrix with diagonal $(a_1, \ldots, a_n)$.

We can show that $\mathbf{T}_n(\alpha_1, \alpha_2)$ is asymptotically normally distributed, provided F is increasing and differentiable in interval $(F^{-1}(\alpha_1) - \varepsilon, F^{-1}(\alpha_2) + \varepsilon)$, and under some regularity conditions imposed on the matrix $\boldsymbol{X}_n$. More precisely,

$$\mathcal{L}\left\{\sqrt{n}(\mathbf{T}_n - \boldsymbol{\beta} - \delta e_1)\right\} \to \mathcal{N}_p\left(\mathbf{0}, \sigma^2 \boldsymbol{Q}^{-1}\right) \qquad (4.61)$$

where $e_1 = (1, 0, \ldots, 0)' \in \mathbb{R}_p$ and

$$\delta = (\alpha_2 - \alpha_1)^{-1} \int_{\alpha_1}^{\alpha_2} F^{-1}(u)du$$

$$\sigma^2 = \sigma^2(\alpha_1, \alpha_2, F) \qquad (4.62)$$

$$= (\alpha_2 - \alpha_1)^{-1}\left\{ \int_{\alpha_1}^{\alpha_2} \alpha_2(F^{-1}(u) - \delta)^2 du \right.$$

$$+ \alpha_1(F^{-1}(\alpha_1) - \delta)^2 + (1 - \alpha_2)(F^{-1}(\alpha_2) - \delta)^2$$

$$\left. - \left[\alpha_1(F^{-1}(\alpha_1) - \delta) + (1 - \alpha_2)(F^{-1}(\alpha_2) - \delta)\right]^2 \right\}$$

In the symmetric situation, when $F(x) + F(-x) = 1$, $x \in \mathbb{R}$ and $\alpha_1 = \alpha$, $\alpha_2 = 1 - \alpha$, $0 < \alpha < \frac{1}{2}$, δ vanishes and $\sqrt{n}(\mathbf{T}_n(\alpha) - \boldsymbol{\beta})$ has asymptotic normal distribution $\mathcal{N}_p(\mathbf{0}, \sigma^2(\alpha, F)\boldsymbol{Q}^{-1})$ with

$$\sigma^2(\alpha, F) = \frac{\int_\alpha^{1-\alpha}(F^{-1}(u))^2 du + 2\alpha(F^{-1}(\alpha))^2}{1 - 2\alpha} \qquad (4.63)$$

Notice that $\sigma^2(\alpha, F)$ coincides with the asymptotic variance of the α-trimmed mean in the location model.

Besides the trimmed least squares estimator we can consider the broad class of *L*-estimators of the form

$$\mathbf{T}_n^\nu = \int_0^1 \widehat{\boldsymbol{\beta}}_n(\alpha)d\nu(\alpha) \qquad (4.64)$$

where ν is a suitable signed measure $(0,1)$ (finite and with a compact support that is a subset of $(0,1)$). The special case is linear combinations of a finite number of regression quantiles that are generated by an atomic measure ν. The trimmed L-estimators, extending the (α_1, α_2)-trimmed mean, are generated by ν that has a Lebesgue measure density

$$J(u) = \frac{I[\alpha_1 \leq u \leq \alpha_2]}{\alpha_2 - \alpha_1}, \quad 0 < \alpha_1 < \alpha_2 < 1$$

Unlike the M-estimators, the L-estimators of the regression parameter are both regression and scale equivariant. Regression quantiles and L-estimators of various types are studied, e.g., in Gutenbrunner (1986), Gutenbrunner and Jurečková (1992), Jurečková and Sen (1996), and recently in Koenker (2005).

The least absolute deviations estimator (regression 0.5-quantile, median regression) is possible to find in the package *quantreg*. Function `rq()` enables computation of the regression quantiles and the function `ranks()` computes the regression rank scores.

```
> library(quantreg)
> data(engel)
> rq(foodexp~income,data=engel,tau=0.5)
Call:
rq(formula = foodexp ~ income, tau = 0.5, data = engel)

Coefficients:
(Intercept)        income
81.4822474    0.5601806

Degrees of freedom: 235 total; 233 residual

> quant<-c(.05,.2,0.5,.8,.95)
> coefficients(rq(foodexp~income,data=engel,tau=quant[-3]))
                tau= 0.05    tau= 0.20   tau= 0.80   tau= 0.95
(Intercept) 124.8800408 102.3138823 58.0066635 64.1039632
income        0.3433611   0.4468995  0.6595106  0.7090685
```

The trimmed least squares estimator we could implement with the help of `rq()` as follows:

```
> "tls.KB"<-function(formula, data,  alpha=0.05)
+ {
+   resid1 <-residuals(rq(formula,data,tau=alpha))
+   resid2 <-residuals(rq(formula,data,tau=1-alpha))
+   c1 <- c(resid1 >= 0)
+   c2 <- c(resid2 <= 0)
+ coefficients(lm(formula,data[c(c1 & c2),]))
+ }
```

```
>
> tls.KB(foodexp~income,data=engel,alpha=0.05)
(Intercept)        income
147.6875780    0.4835946
> tls.KB(foodexp~income,data=engel,alpha=0.2)
(Intercept)        income
95.5841657    0.5421883
```

With the following code we can display (see Figure 4.8) the selected regression quantile lines, the trimmed least squares line and the least squares line.

```
> plot(foodexp~income,data=engel)
> for (i in quant)
+ abline(rq(foodexp~income,data=engel,tau=i))
> abline(tls.KB(foodexp~income,data=engel,alpha=0.2),lty=2)
> abline(lm(foodexp~income,data=engel),lty=3)
>
```

4.7 Regression rank scores

The dual program to (4.56) has a very interesting interpretation: while the solutions of (4.56) are the regression quantiles, the solutions of the dual problem are called *regression rank scores* because they remind the ranks of observations and have many properties similar to those of ranks.

The dual program to (4.56) can be written in the form

$$\sum_{i=1}^{n} Y_i \hat{a}_i := \max$$

$$\text{under constraint} \quad \sum_{i=1}^{n} \hat{a}_i = n(1-\alpha) \tag{4.65}$$

$$\sum_{i=1}^{n} x_{ij} \hat{a}_i = (1-\alpha) \sum_{i=1}^{n} x_{ij}, \; j = 2, \ldots, p$$

$$0 \le \hat{a}_i \le 1, \quad i = 1, \ldots, n, \quad 0 < \alpha < 1$$

The components of the optimal solution of (4.65),

$$\hat{\mathbf{a}}_n(\alpha) = (\hat{a}_{n1}(\alpha), \ldots, \hat{a}_{nn}(\alpha))', \quad 0 \le \alpha \le 1$$

are called the *regression rank scores*. The matrix form of program (4.65) is more compact:

$$\mathbf{Y}'_n \hat{\mathbf{a}} := \max$$

$$\text{under constraint} \quad \mathbf{X}'_n \hat{\mathbf{a}} = (1-\alpha) \mathbf{X}'_n \mathbf{1}_n \tag{4.66}$$

$$\hat{\mathbf{a}} \in [0, 1]^n, \; 0 \le \alpha \le 1$$

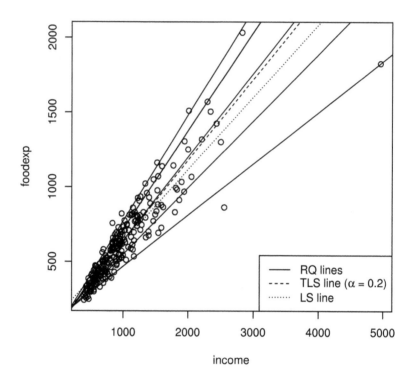

Figure 4.8 *Plot of regression α-quantile lines ($\alpha = 0.05, 0.2, 0.5, 0.8, 0.95$), trimmed least squares line ($\alpha = 0.2$) and least square line for Engel food expenditure data.*

Recall that $x_{i1} = 1$, $i = 1, \ldots, n$ by assumption (4.53). From (4.66) we can easily verify that the regression rank scores are invariant with respect to changes of $\boldsymbol{\beta}$, i.e.,

$$\widehat{\mathbf{a}}_n(\alpha, \boldsymbol{Y} + \boldsymbol{Xb}) = \widehat{\mathbf{a}}_n(\alpha, \boldsymbol{Y}) \quad \forall \boldsymbol{b} \in \mathbb{R}_p \tag{4.67}$$

Because $\widehat{\boldsymbol{\beta}}(\alpha)$ and $\widehat{\mathbf{a}}_n(\alpha)$ are dual to each other, we get from the linear programming theory

$$\hat{a}_{ni}(\alpha) = \begin{cases} 1 & \ldots \quad Y_i > \boldsymbol{x}_i'\widehat{\boldsymbol{\beta}}_n(\alpha) \\ 0 & \ldots \quad Y_i < \boldsymbol{x}_i'\widehat{\boldsymbol{\beta}}_n(\alpha), \quad i = 1, \ldots, n \end{cases} \tag{4.68}$$

and if $Y_i = \boldsymbol{x}_i'\widehat{\boldsymbol{\beta}}_n(\alpha)$ for some i (the exact fit), then $0 < \hat{a}_{ni}(\alpha) < 1$; there are exactly p such components for each α, and the corresponding values of $\hat{a}_{ni}(\alpha)$ are determined by the restriction conditions in (4.66). The regression rank scores $\hat{a}_{ni}(\alpha)$, $i = 1, \ldots, n$ are continuous, piecewise linear functions of $\alpha \in [0, 1]$ satisfying $\hat{a}_{ni}(0) = 1$, $\hat{a}_{ni}(1) = 0$ (see Figure 4.9).

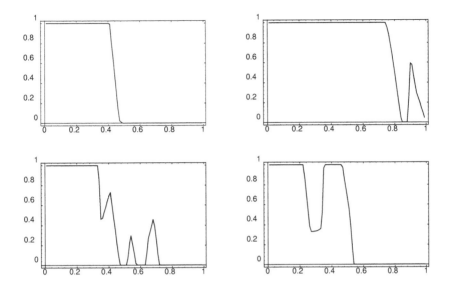

Figure 4.9 *Example of regression rank scores for the simulated data.*

Due to these nice properties, the regression rank scores have many applications. The invariance property (4.67) guarantees that the tests based on regression rank scores are invariant to $\boldsymbol{\beta}$ in the situations when $\boldsymbol{\beta}$ is a nuisance parameter, while another parameter is of interest for testing a hypothesis or estimation, or when we want to test a hypothesis on the shape of the distribution. The fact that in this way we avoid an estimation of $\boldsymbol{\beta}$ not only facilitates the computation, but we also prevent a risk of a wrong choice of an estimator of $\boldsymbol{\beta}$.

The tests of linear hypotheses with nuisance parameter $\boldsymbol{\beta}$, based on regression rank scores, were constructed by Gutenbrunner et al. (1993). The tests based on regression rank scores enable verification of hypotheses of the type $\mathbf{H}: \boldsymbol{\beta}^{(2)} = \mathbf{0}$ in the model

$$\mathbf{Y} = \mathbf{X}^{(1)}\boldsymbol{\beta}^{(1)} + \mathbf{X}_2\boldsymbol{\beta}^{(2)} + \mathbf{e} \tag{4.69}$$

with the nuisance parameter $\boldsymbol{\beta}^{(1)}$, that contains an intercept. Under $\mathbf{H}$, the model (4.69) reduces to the form

$$\mathbf{Y} = \mathbf{X}^{(1)}\boldsymbol{\beta}^{(1)} + \mathbf{e} \tag{4.70}$$

and the test is based on the regression rank scores corresponding to the hypothetical model (4.70). The tests have the same asymptotic powers as the rank tests of $\mathbf{H}$ in the situation that $\boldsymbol{\beta}^{(1)}$ is fully known. The exact forms of the tests and more details can be found in Gutenbrunner et al. (1993).

This test is implemented in the *quantreg* package and can be used as follows.

```
> library(quantreg)
> data(engel)
> fit1<-rq(foodexp~income,data=engel)
> fit0<-rq(foodexp~1,data=engel)
> anova(fit1,fit0, test="rank")
Quantile Regression Analysis of Deviance Table

Model 1: foodexp ~ income
Model 2: foodexp ~ 1
Df Resid Df F value     Pr(>F)
1  1        233  91.127 < 2.2e-16 ***
---
Signif. codes:  0 '***' 0.001 '**' 0.01 '*' 0.05 '.' 0.1 ' ' 1
>
```

As we have already mentioned in Section 4.3, a scale equivariant M-estimator of the regression parameter can be obtained using studentization by a suitable scale statistic. However, such statistics should be scale equivariant and invariant to the change of the regression parameter (see 4.29), and such scale statistics are based on regression rank scores and on regression quantiles. We shall describe some of these scale statistics in the next section.

4.8   Robust scale statistics

The M-estimators of regression parameters are regression equivariant but not scale-equivariant; hence either they should be studentized, or the scale should be estimated simultaneously with the regression parameters. The studentizing scale statistic $S_n(Y)$ should be fully determined by the observations Y_i and by matrix $\mathbf{X}_n$ and it should satisfy

$$S_n(c(\boldsymbol{Y} + \boldsymbol{X}\boldsymbol{b})) = cS_n(\boldsymbol{Y}) \quad \forall \boldsymbol{b} \in \mathbb{R}_p, \ c > 0, \ \boldsymbol{Y} \in \mathbb{R}_n \qquad (4.71)$$

regression invariance and scale equivariance, cf. (4.29).

There are not many statistics of this type in the literature; when we studentize an M-estimator by S_n that is only invariant to the shift, but not to the regression; then the M-estimator loses its regression equivariance. We shall describe some statistics satisfying (4.71), based on regression quantiles or on the regression rank scores.

(i) *Median absolute deviation from the regression median* (MAD).

Statistic MAD is frequently used in the location model. Welsh (1986) extended MAD to the linear regression model in the following way: start with an initial estimator β^0 of β, $\sqrt{n}$-consistent, regression and scale equivariant (so we shall not start with the ordinary M-estimator). The Welsh scale statistic is then defined as

$$S_n = \text{med}_{1 \leq i \leq n} \left| Y_i(\boldsymbol{\beta}^0) - \xi_{\frac{1}{2}}(\boldsymbol{\beta}^0) \right| \qquad (4.72)$$

where

$$Y_i(\boldsymbol{\beta}^0) = Y_i - \mathbf{x}_i'\boldsymbol{\beta}^0, \quad i = 1, \ldots, n$$

$$\xi_{\frac{1}{2}}(\boldsymbol{\beta}^0) = \mathrm{med}_{1 \le i \le n} Y_i(\boldsymbol{\beta}^0)$$

S_n is apparently invariant/equivariant in the sense of (4.71). Its asymptotic properties are studied in the Welsh (1986) paper.

(ii) *L-statistics based on regression quantiles.*
The Euclidean distance of two regression quantiles

$$S_n = \left\| \widehat{\boldsymbol{\beta}}_n(\alpha_2) - \widehat{\boldsymbol{\beta}}_n(\alpha_1) \right\| \tag{4.73}$$

$0 < \alpha_1 < \alpha_2 < 1$, is invariant/equivariant in the sense of (4.71) and $S_n \xrightarrow{P} S(F) = F^{-1}(\alpha_2) - F^{-1}(\alpha_1)$ as $n \to \infty$. The Euclidean norm can be replaced by L_p-norm or another suitable norm. It is also possible to use only the absolute difference of the first components of the regression quantiles, i.e.,

$$S_n = \left| \widehat{\beta}_{n1}(\alpha_2) - \widehat{\beta}_{n1}(\alpha_1) \right|$$

(iii) Bickel and Lehmann (1979) proposed several *measures of spread* of distribution F, e.g.,

$$S(F) = \left\{ \int_{\frac{1}{2}}^1 \left[F^{-1}(u) - F^{-1}(1-u) \right]^2 d\Lambda(u) \right\}^{\frac{1}{2}}$$

where Λ is the uniform probability distribution on interval $(\frac{1}{2}, 1 - \delta)$, $0 < \delta < \frac{1}{2}$. Using their idea, we obtain a broad class of scale statistics based on regression quantiles of the following type

$$S_n = \left\{ \int_{\frac{1}{2}}^1 \left\| \widehat{\boldsymbol{\beta}}_n(u) - \widehat{\boldsymbol{\beta}}_n(1-u) \right\|^2 d\Lambda(u) \right\}^{\frac{1}{2}}$$

Again, the squared norm $\left\| \widehat{\boldsymbol{\beta}}_n(u) - \widehat{\boldsymbol{\beta}}_n(1-u) \right\|^2$ in the integral can be replaced by the squared difference of the first components of $\widehat{\boldsymbol{\beta}}_n(u)$ and $\widehat{\boldsymbol{\beta}}_n(1-u)$.

(iv) *Estimators of $1/f(F^{-1}(\alpha))$ based on regression quantiles.*
The density quantile function $f(F^{-1}(\alpha))$, $0 < \alpha < 1$ is an important characteristic of the probability distribution, necessary in the statistical inference based on quantiles. Unfortunately, it is not easy to estimate $f(F^{-1}(\alpha))$, even if we only need its value at a single α. Similar to when we

estimate the density, we are able to estimate $f(F^{-1}(\alpha))$ only with a lower rate of consistency than the usual $\sqrt{n}$.

Falk (1986) constructed estimates of $f(F^{-1}(\alpha))$ for the location model, based on the sample quantiles. These estimates are either of the histogram or the kernel type. Benefiting from the convenient properties of regression quantiles, we can extend Falk's estimates to the linear regression model, replacing the sample quantiles with the intercept components of the regression quantiles. The histogram type estimator has the form

$$H_n^{(\alpha)} = \frac{1}{2\nu_n} \left[\widehat{\beta}_{n1}(\alpha + \nu_n) - \widehat{\beta}_{n1}(\alpha - \nu_n) \right] \tag{4.74}$$

where the sequence $\{\nu_n\}$ is our choice and it should satisfy

$$\nu_n = o\left(n^{-\frac{1}{2}}\right) \quad \text{and} \quad \lim_{n \to \infty} n\nu_n = \infty$$

The kernel type estimator of $1/f(F^{-1}(\alpha))$ is based on the kernel function $k : \mathbb{R}_1 \mapsto \mathbb{R}_1$ that has a compact support and satisfies the relations

$$\int k(x)dx = 0 \quad \text{and} \quad \int xk(x)dx = -1$$

The kernel estimator then has the form

$$\chi_n^{(\alpha)} = \frac{1}{\nu_n^2} \int_0^1 \widehat{\beta}_{n1}(u)k\left(\frac{\alpha - u}{\nu_n}\right) du \tag{4.75}$$

where this time the sequence $\{\nu_n\}$ should satisfy

$$\nu_n \to 0, \quad n\nu_n^2 \to \infty, \quad n\nu_n^3 \to 0, \quad \text{as } n \to \infty$$

Both (4.74) and (4.75) are $\sqrt{n\nu_n}$-consistent estimators of $1/f(F^{-1}(\alpha))$ and are invariant/equivariant in the sense of (4.71). Their lower rate of consistency is analogous to that of the density estimator and cannot be considerably improved. As such, they are not usually used for a studentization, but they are very useful in various contexts, e.g., in the inference on the quantiles of F. Their asymptotic properties, applications and numerical illustrations are studied in detail in the book of Dodge and Jurečková (2000).

(v) *Scale statistics based on the regression rank scores.*

Let $(\hat{a}_{n1}(\alpha), \ldots, \hat{a}_{nn}(\alpha))$, $0 < \alpha < 1$ be the regression rank scores for model (4.1). Choose a nondecreasing score function $\varphi : (0,1) \mapsto \mathbb{R}_1$ standardized so that

$$\int_{\alpha_0}^{1-\alpha_0} \varphi^2(\alpha)d\alpha = 1$$

for a fixed α_0, $0 < \alpha_0 < \frac{1}{2}$ and calculate the scores

$$\hat{b}_{ni} = -\int_{\alpha_0}^{1-\alpha_0} \varphi(\alpha)d\hat{a}_{ni}(\alpha), \quad i = 1, \ldots, n$$

The scale statistic

$$S_n = \frac{1}{n} \sum_{i=1}^{n} Y_i \hat{b}_{ni} \tag{4.76}$$

is invariant/equivariant in the sense of (4.71) and it is an $\sqrt{n}$-consistent estimator of the functional

$$S(F) = \int_{\alpha_0}^{1-\alpha_0} \varphi(\alpha) F^{-1}(\alpha) d\alpha$$

4.9   Estimators with high breakdown points

The breakdown point of an estimator in the linear model takes into account not only possible replacements of observations $Y_1, \ldots, Y_n$ by arbitrary values, but also possible replacements of vectors $(\boldsymbol{x}_1', Y_1)', \ldots, (\boldsymbol{x}_n', Y_n)'$. More precisely, our observations create a matrix

$$\boldsymbol{Z} = \begin{bmatrix} \boldsymbol{z}_1' \\ \boldsymbol{z}_2' \\ \ldots \\ \boldsymbol{z}_n' \end{bmatrix} = \begin{bmatrix} \boldsymbol{x}_1', & y_1 \\ \boldsymbol{x}_2', & y_2 \\ \ldots \\ \boldsymbol{x}_n', & y_n \end{bmatrix}$$

and the breakdown point of estimator $\boldsymbol{T}$ parameter $\boldsymbol{\beta}$ is the smallest integer $m_n(\boldsymbol{Z})$ such that, replacing arbitrary m rows in matrix $\boldsymbol{Z}$ by arbitrary rows and denoting the resulting estimator $\boldsymbol{T}_m^*$, then $\sup \| \boldsymbol{T} - \boldsymbol{T}_m^* \| = \infty$, where the supremum is taken over all possible replacements of m rows. We also often measure the breakdown point by means of a limit $\varepsilon^* = \lim_{n \to \infty} \frac{m_n}{n}$, if the limit exists. We immediately see that even the estimators that have the breakdown point $1/2$ in the location model can hardly attain $1/2$ in the regression model, because the matrix $\boldsymbol{X}$ plays a substantial role. Then one naturally poses questions, e.g., whether there are any estimators with maximal possible breakdown points in the regression model; and if so, what do they look like, is it easy to compute them, and in which context are they useful and desirable?

Siegel replied affirmatively to the first question in 1982, when he constructed a so-called *repeated median* with the 50% breakdown point. In the simple regression model $Y_i = \alpha + x_i \beta + e_i$, $i = 1, \ldots, n$, the repeated median of the slope parameter has a simple form

$$\widehat{\beta} = \mathrm{med}_{1 \le i \le n} \mathrm{med}_{j:j \ne i} \frac{Y_i - Y_j}{x_i - x_j}$$

The repeated median can be calculated using a function from the package *mblm*:

```
> library(quantreg); data(engel)
> library(mblm)
> mblm(foodexp~income,data=engel)

Call:
mblm(formula = foodexp ~ income, dataframe = engel)

Coefficients:
(Intercept)          income
64.0692            0.5966

>
```

However, in the p-dimensional regression, its computation needs $O(n^p)$ operations, hence it is not convenient for practical applications.

Hampel (1975) expressed the idea of the *least median of squares* (LMS) that minimizes

$$\text{med}_{1 \leq i \leq n}\{[Y_i - \boldsymbol{x}_i'\mathbf{t}]^2\}, \ \mathbf{t} \in \mathbb{R}_p \tag{4.77}$$

Rousseeuw (1984) demonstrated that this estimator had the 50% breakdown point. It estimates $\boldsymbol{\beta}$ consistently, but with the rate of consistency $n^{\frac{1}{3}}$ only (Kim and Pollard (1990), Davies (1990)), hence it is highly inefficient. Rousseeuw also introduced the *least trimmed squares estimator* (LTS) as a solution of the minimization (we denote $(r^2)_i = (Y_i - \boldsymbol{x}_i'\mathbf{t})^2$, $i = 1, \ldots, n$):

$$\sum_{i=1}^{h_n} (r^2)_{n:i} := \min, \ \mathbf{t} \in \mathbb{R}_p$$

where $h_n = [n/2] + [(p+1)/2]$ and $[a]$ denotes the integer part of a. This estimator is already an $\sqrt{n}$-consistent estimator of $\boldsymbol{\beta}$ and has the breakdown point $1/2$. Its weighted version was studied by Víšek (2002a, b).

The *least trimmed sum of absolute deviations* (LTA) is another estimator, originally proposed by Bassett (1991), and further studied by Tableman (1994 a, b), Hössjer (1994) and Hawkins and Olive (1999).

To have an estimator with a high breakdown point is desirable in regression and other models, but the high breakdown point is only one advantage that cannot be emphasized above others. Using these estimators in practice, we should be aware of the possibility that, being resistant to outliers and gross errors, they can be sensitive to small perturbations in the central part of the data. We can refer to to Hettmansperger and Sheather (1992) and to Ellis (1998) for numerical evidence of this problem, while it still needs a thorough analytical treatment.

Package MASS also contains the procedure lqs for a regression estimator with a high breakdown point. We can select either the LMS estimator or the LTS estimator. The procedure uses an approximate resampling algorithm, so the minimization is only approximate. If n is not too large, the number of samples can be increased by setting a parameter `nsam="exact"` to get a lower objective function.

```
> library(MASS)
> set.seed(2019); library(quantreg); data(engel)
> lqs(foodexp~income,data=engel, method = "lms")
Call:
lqs.formula(formula = foodexp ~ income, data = engel,
    method = "lms")

Coefficients:
(Intercept)          income
-3.5997          0.6909

Scale estimates 63.48 63.86

> lqs(foodexp~income,data=engel, method = "lts")
Call:
lqs.formula(formula = foodexp ~ income, data = engel,
            method = "lts")

Coefficients:
(Intercept)          income
-23.9090          0.7287

Scale estimates 70.89 68.94

> (lqs.mod <- lqs(foodexp~income,data=engel, method = "lts",
+                 nsam="exact"))
Call:
lqs.formula(formula = foodexp ~ income, data = engel,
            nsam = "exact", method = "lts")

Coefficients:
(Intercept)          income
-23.8185          0.7286

Scale estimates 70.89 69.88

>
```

We could find a function `ltsReg` for LTS regression also in the package *robustbase*. See also a function `ga.lts` for LTS regression using genetic algorithms in the package *galts*. We can compare estimates obtained by functions `lts` and `ga.lts` according to their criterion value:

```
> set.seed(123); library(galts)
> ga.lts(engel$foodexp~engel$income, lower = -30, upper = 3,
+         iters = 80,csteps=4)
$'coefficients'
[1] -23.8353671    0.7286363

$crit
[1] 85381.77

$method
[1] "ga"

> sum(sort(resid(lqs.mod)^2)[1:118])
[1] 85381.77
>
```

4.10 S-estimators and MM-estimators

Other high breakdown point estimators are the S-estimator, MM-estimators
and τ estimators (see Rousseeuw and Yohai (1984)), whose structures were
already described in Section 3.6 on the location model. These estimators are
studied in detail in the book by Rousseeuw and Leroy (1987); great attention
is also devoted to their computational aspects. These estimators combine the
high breakdown point with the proper rate of consistency $n^{\frac{1}{2}}$, and also with
a fairly good efficiency at the normal model. The S-estimator, proposed by
Rousseeuw and Yohai (1984), minimizes an estimator of scale,

$$\tilde{\beta}_n = \arg \min \ \tilde{S}_n(\mathbf{b})$$

and the estimator of scale $\tilde{S}_n(\mathbf{b})$ solves the equation

$$\frac{1}{n} \sum_{i=1}^{n} \rho \left(\frac{Y_i - \mathbf{x}_i^\top \beta}{\tilde{S}_n(\beta)} \right) = \gamma \ \text{ for each fixed } \ \beta$$

where the function $\rho(x)$ satisfies conditions (i)–(iv) in Section 3.6. Following
the steps of the proof of Lemma 3.1, we can replace the minimization of $\tilde{S}_n(\mathbf{b})$
with minimization of the sum of absolute values of residuals. Then it follows
from Theorem 1 of Rousseeuw (1982) that the estimator $\tilde{\beta}$ has the breakdown
point

$$\varepsilon_n^* = \frac{1}{n} \left(\left[\frac{n}{2} \right] - p + 2 \right)$$

provided the observations $((\mathbf{x}_1, y_1), \dots, (\mathbf{x}_1, y_1))$ are in *general position*.

Remark 4.2 *We say that the observations are in general position when any
p of them give a unique determination of β. In the case of simple regression
($p = 2$) , this means that no two points coincide or determine a vertical line.*

For computing the *MM*-estimator the function `rlm` from *MASS* package can be used as in the previous chapter.

```
> library(quantreg); data(engel)
> library(MASS)
> rlm(foodexp~income,data=engel,method="MM")
Call:
rlm(formula = foodexp ~ income, data = engel, method = "MM")
Converged in 15 iterations

Coefficients:
(Intercept)      income
85.5072673    0.5539062

Degrees of freedom: 235 total; 233 residual
Scale estimate: 76.3
> summary(rlm(foodexp~income,data=engel,method="MM"))

Call: rlm(formula = foodexp ~ income, data = engel,
  method = "MM")
Residuals:
Min        1Q     Median       3Q        Max
-1004.471   -55.202     1.173    51.141    383.753

Coefficients:
Value    Std. Error t value
(Intercept) 85.5072 12.0796     7.0787
income        0.5539  0.0109    50.9320

Residual standard error: 76.31 on 233 degrees of freedom
> deviance(rlm(foodexp~income,data=engel,method="MM"))
[1] 3339046
>
```

Also the function `lmrob` from the package *robustbase* can be again employed:

```
library(robustbase)
> (MM <- lmrob(foodexp~income,data=engel))

Call:
lmrob(formula = foodexp ~ income, data = engel)
\--> method = "MM"
Coefficients:
(Intercept)       income
85.5235        0.5539
```

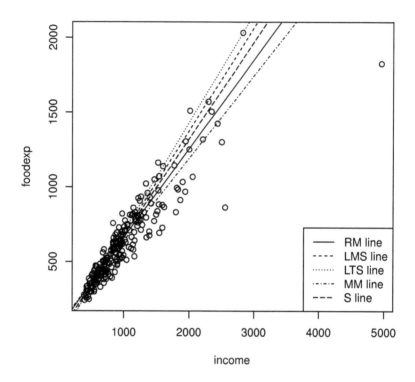

Figure 4.10 *Plot of repeated median, least median of squares, least trimmed squares, MM-estimate and S-estimate lines for Engel food expenditure data.*

```
> MM$init.S$coefficients
(Intercept)        income
27.2533369    0.6482646
> lmrob.S(x = model.matrix(foodexp~income,data=engel),
+   engel$foodexp, control = lmrob.control())$coefficients
(Intercept)        income
27.2533565    0.6482646
>
```

The following code generates Figure 4.10 with estimates of Sections 4.9 and 4.10.

```
> plot(foodexp~income,data=engel)
> abline(mblm(foodexp~income,data=engel))
> abline(lqs(foodexp~income,data=engel, method = "lms"),lty=2)
> abline(lqs.mod,lty=3)
> abline(MM,lty=4)
```

```
> abline(MM$init.S,lty=5)
> legend("bottomright",c("RM line", "LMS line","LTS line",
+                        "MM line", "S line" ),lty = 1:5)
>
```

4.11   Examples

Example 4.2 *Consider the dataset used by Pearson in 1906 to study the relationship between a parent's and his or her child's height. The dataset (see Figure 4.11) contains 1078 measurements of a father's height and his son's height in inches, and it was used by Pearson to investigate the regression. Data are available from R package, UsingR, dataset* **father.son**.

Table 4.1 *Results for Pearson data.*

Methods used for estimation	$\hat{\beta}_0$	$\hat{\beta}_1$
Least square	33.8866	0.5141
M-estimation	34.5813	0.5039
MM-estimation	34.7037	0.5021
Least absolute deviation	35.5710	0.4885
Trimmed least squares ($\alpha = 0.05$)	34.3876	0.5068
Trimmed least squares ($\alpha = 0.20$)	35.0088	0.4974
Least median of squares	45.9288	0.3281
Least trimmed squares	38.7060	0.4391
Repeated median	33.7523	0.5174

We have the model

$$Y_i = \beta_0 + \beta_1 X_{1i} + e_i, \quad i = 1, \ldots, 1078$$

where Y is the son's height, X_1 is the father's height. We want to determine the unknown values β_0 and β_1, assuming that the errors $e_1, \ldots, e_{1078}$ are independent and identically distributed, symmetrically around zero.

In Table 4.1 we will now give the results obtained by the different methods of regression used in the preceding sections.

We see that the values of estimated coefficients are rather close to each other with the exception of the least median of squares estimation. Figure 4.11 illustrates this difference, i.e., least squares regression and least median of squares lines.

Example 4.3 *Another example of simple regression is from astronomy: the Hertzsprung-Russell diagram of the Star Cluster CYG OB1, which contains 47 stars in the direction of Cygnus. The first variable is the logarithm of the effective temperature at the surface of the star (X) and the second one is the logarithm of its light intensity (Y). This dataset was analyzed by Rousseeuw*

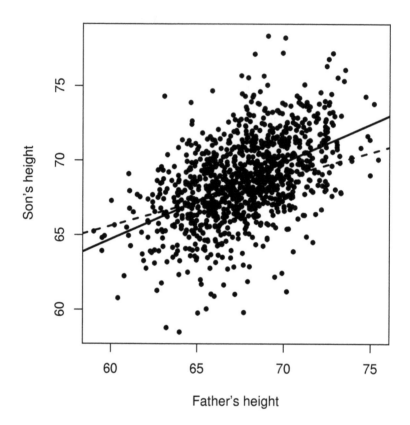

Figure 4.11 *Least squares (solid line) and least median of squares (dashed line) regression for the Pearson dataset contains 1078 measurements of a father's and his son's height in inches.*

and Leroy (1987), who compared the least squares method with the least median of squares estimator. The data contains the four giant stars — outliers, which are leverage points but they are not errors (see dataset **starsCYG** in package robustbase).

We again have the model

$$Y_i = \beta_0 + \beta_1 X_{1i} + e_i, \quad i = 1, \ldots, 47$$

where Y is the log light intensity and X_1 is the log of the temperature. The results obtained by different methods are summarized in Table 4.2.

We see that not only the least square regression but also M, trimmed least squares and least absolute deviation regression, are sensitive to a leverage

Table 4.2 *Results for data of the Hertzsprung-Russell diagram of the Star Cluster CYG OB1.*

Methods used for estimation	$\hat{\beta}_0$	$\hat{\beta}_1$
Least square	6.7935	−0.4133
M-estimation	6.8658	−0.4285
MM-estimation	−4.9702	2.2533
Least absolute deviation	8.1492	−0.6931
Trimmed least squares ($\alpha = 0.05$)	6.8361	−0.4134
Trimmed least squares ($\alpha = 0.20$)	6.9424	−0.4459
Least median of squares	−12.76	4.00
Least trimmed squares	−14.05	4.32
Repeated median	−6.065	2.500

point. Figure 4.12 displays the data and the least squares, least absolute deviation, repeated median, least trimmed squares and MM-estimate lines.

Example 4.4 *As another example we consider the Brownlee's stack loss plant data, see* stackloss *dataset in the package datasets. The dataset consists of one dependent variable (stack loss) and three independent variables, see Figure 4.13. The aim is to estimate the parameters of the following multiple linear regression model*

$$Y_i = \beta_0 + \beta_1 X_{1i} + \beta_1 X_{2i} + \beta_1 X_{3i} + e_i, \quad i = 1, \ldots, 21$$

where Y is the stack loss, X_1 is the flow of cooling air, X_2 is the cooling water inlet temperature and X_3 is the concentration of acid. We want to determine the unknown values β_0, β_1, β_2 and β_3, assuming that the errors $e_1, \ldots, e_{21}$ are independent and identically distributed, symmetrically around zero.

For the parameters of this model we can obtain by the following code the least square estimate, M-estimate, MM-estimate, least absolute deviation estimate, trimmed least squares estimates, least median of squares and least trimmed squares estimator.

```
> library(MASS); library(quantreg)
> plot(stackloss)
> lm(stack.loss ~ .,data=stackloss)

Call:
lm(formula = stack.loss ~ ., data = stackloss)

Coefficients:
(Intercept)      Air.Flow     Water.Temp      Acid.Conc.
  -39.9197        0.7156         1.2953         -0.1521

>
```

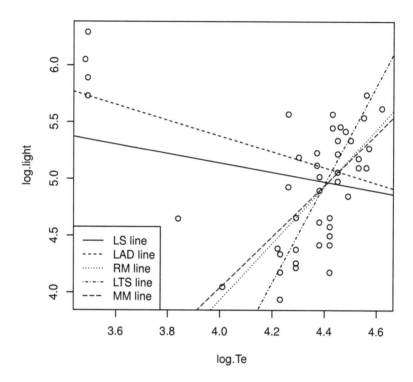

Figure 4.12 *Plot of least squares, least absolute deviation, repeated median, least trimmed squares and MM-estimate lines for the dataset of the Hertzsprung-Russell diagram of the Star Cluster CYG OB1.*

```
> rlm(stack.loss ~ .,data=stackloss)
Call:
rlm(formula = stack.loss ~ ., data = stackloss)
Converged in 9 iterations

Coefficients:
(Intercept)    Air.Flow   Water.Temp   Acid.Conc.
-41.0265311   0.8293739    0.9261082   -0.1278492

Degrees of freedom: 21 total; 17 residual
Scale estimate: 2.44
>
> rlm(stack.loss ~ .,data=stackloss, method="MM")
```

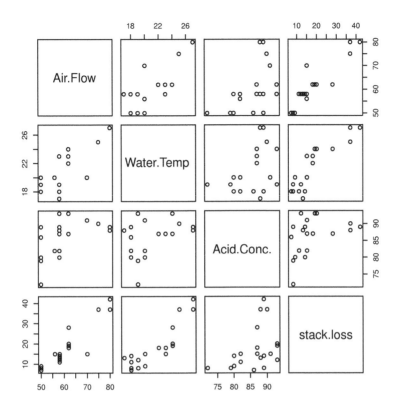

Figure 4.13 *Plot of* 21 *observations for the stack loss plant dataset.*

```
Call:
rlm(formula = stack.loss ~ ., data = stackloss, method = "MM")
Converged in 12 iterations

Coefficients:
(Intercept)      Air.Flow   Water.Temp    Acid.Conc.
-41.7072704     0.9372710    0.5940633    -0.1129477

Degrees of freedom: 21 total; 17 residual
Scale estimate: 1.98
>
>
> rq(stack.loss ~ .,data=stackloss)
Call:
rq(formula = stack.loss ~ ., data = stackloss)
```

```
Coefficients:
(Intercept)      Air.Flow    Water.Temp    Acid.Conc.
-39.68985507    0.83188406    0.57391304    -0.06086957

Degrees of freedom: 21 total; 17 residual
>
> tls.KB(stack.loss ~ .,data=stackloss,alpha=0.05)
(Intercept)      Air.Flow    Water.Temp    Acid.Conc.
-41.1558281     0.8443507     0.9525682     -0.1364353
>
> tls.KB(stack.loss ~ .,data=stackloss,alpha=0.2)
(Intercept)      Air.Flow    Water.Temp    Acid.Conc.
-42.4723383     0.9671259     0.6124727     -0.1318105
>
> lqs(stack.loss ~ .,data=stackloss, method = "lms",
+      nsamp="exact")
Call:
lqs.formula(formula = stack.loss ~ ., data = stackloss,
 nsamp = "exact", method = "lms")

Coefficients:
(Intercept)      Air.Flow    Water.Temp    Acid.Conc.
-34.2500         0.7143        0.3571        0.0000

Scale estimates 0.5514 0.4798

> lqs(stack.loss ~ .,data=stackloss, method = "lts",
+      nsamp="exact")
Call:
lqs.formula(formula = stack.loss ~ ., data = stackloss,
 nsamp = "exact", method = "lts")

Coefficients:
(Intercept)      Air.Flow    Water.Temp    Acid.Conc.
-3.581e+01      7.500e-01     3.333e-01     3.489e-17

Scale estimates 0.8482 0.8645

>
```

4.12   Problems and complements

4.1 The number of distinct solutions of the linear program (4.56), as α runs from 0 to 1, is of order $O_p(n \cdot \ln n)$. In the location model, when $\mathbf{X} = \mathbf{1}_n$, the number of distinct values is exactly 1 (Portnoy (1991)).

4.2 The L_1-estimator and any other M-estimator of $\boldsymbol{\beta}$ in model (4.1) has the breakdown point $\frac{1}{n}$.

4.3 Another possible characteristic of performance of L-estimators in the linear model is the largest integer m such that, for any set $M \subset N = \{1, \ldots, n\}$ of size m,

$$\inf_{\|\mathbf{b}\|=1} \left\{ \frac{\sum_{i \in N-M} |\mathbf{x}_i' \mathbf{b}|}{\sum_{i \in N} |\mathbf{x}_i' \mathbf{b}|} \right\} > \frac{1}{2}$$

4.4 The restricted M-estimator is a solution either of minimization (4.25) or of minimization (4.28) under the linear constraint

$$\mathbf{A}\boldsymbol{\beta} = \mathbf{c} \tag{4.78}$$

where $\mathbf{A}$ is a $q \times p$ matrix of full rank and $\mathbf{c} \in \mathbb{R}^p$. The relation (4.78) can be our hypothesis and we are interested in the behavior of the estimator in the hypothetical situation.

4.5 Let $r_i = Y_i - \mathbf{x}_i' \widehat{\boldsymbol{\beta}}_0$, $i = 1, \ldots, n$ be the residuals with respect to estimator $\widehat{\boldsymbol{\beta}}_0$ of $\boldsymbol{\beta}$, and let θ_n be the solution of the minimization

$$\sum_{i=1}^{n} \rho_\alpha(r_i - \theta_n) := \min$$

where ρ_α is the criterion function of an α-regression quantile defined in (4.55), $0 < \alpha < 0$, then

$$n^{-\frac{1}{2}} \sum_{i=1}^{n} \psi_\alpha(r_i - \theta_n) \to 0 \text{ a.s. as } n \to \infty$$

where $\psi_\alpha(x) = \alpha - I[x < 0]$, $x \in \mathbb{R}$ (Ruppert and Carroll (1980)).

4.6 Let $\widehat{\boldsymbol{\beta}}_n(\alpha)$ be a solution of the minimization (4.54) (the regression α-quantile, then

$$n^{-\frac{1}{2}} \sum_{i=1}^{n} \mathbf{x_i} \psi_\alpha \left(Y_i - \mathbf{x}_i' \widehat{\boldsymbol{\beta}}_n(\alpha) \right) \to 0 \text{ a.s. as } n \to \infty$$

where ψ_α is the function from problem **4.4** (Ruppert and Carroll (1980)).

4.7 The Wilcoxon-type test of hypothesis $\mathbf{H} : \boldsymbol{\beta}^{(2)} = \mathbf{0}$ in the model

$$\mathbf{Y} = \mathbf{X}^{(1)} \boldsymbol{\beta}^{(1)} + \mathbf{X}_2 \boldsymbol{\beta}^{(2)} + \mathbf{e}$$

(see (4.69), $\mathbf{X}^{(1)}$ is $n \times p$ and $\mathbf{X}^{(2)}$ is $n \times q$), based on the regression rank scores, is based on the following criterion. Assume that $\boldsymbol{\beta}^{(1)}$ contains an intercept, i.e., the first column of $\mathbf{X}^{(1)}$ is $\mathbf{1}_n$. Calculate the regression rank scores $(\hat{a}_{n1}(\alpha), \ldots, \hat{a}_{nn}(\alpha))$, $0 \leq \alpha \leq 1$, for the hypothetical model $\mathbf{Y} = \mathbf{X}^{(1)}\boldsymbol{\beta}^{(1)} + \mathbf{e}$, the Wilcoxon scores

$$\hat{b}_n = \int_0^1 u \, d\hat{a}_{ni}, \quad i = 1, \ldots, n$$

and linear rank statistics vector $\mathbf{S}_n = \sum_{i=1}^n \mathbf{x}_i^{(2)} \hat{b}_i$. Then we reject $\mathbf{H}$ provided

$$T_n = \frac{1}{12} \mathbf{S}_n' \mathbf{Q}_n^{-1} \mathbf{S}_n \geq \chi_q^2(.95)$$

where $\chi_q^2(.95)$ is the 95% quantile of the χ_q^2 distribution (Gutenbrunner et al. (1993)).

4.8 Apply the described methods on the "salinity" dataset consisting of measurements of water salinity (salt concentration, lagged salinity, trend) and river discharge taken in North Carolina. This dataset was listed by Ruppert and Carroll (1980) and the package *rrcov* contains it.

4.9 Let us consider an R-estimator minimizing the rank dispersion (4.48) with the scores $a_n(i) = n \int_{(i-1)/n}^{i/n} \varphi(u) du$, where

$$\varphi_\lambda(u) = \lambda - I[u < \lambda], \quad 0 < u < 1, \quad \lambda \in (0, 1) \text{ fixed.}$$

If we set $\lambda = \alpha$ then this R-estimator coincides with the slope components of regression α-quantile $\widehat{\boldsymbol{\beta}}(\alpha)$ solving (4.54), see Jurečková (2017) and Jurečková et. al. (2017).

For arbitrary fixed $\lambda = \alpha$, compute and compare both estimates (the R-estimator and the slope components of $\widehat{\boldsymbol{\beta}}(\alpha)$) for the Engel food expenditure data (dataset `engel` in package *quantreg*) and for the stackloss dataset from Example 4.4. For the computation of the R-estimate with the specified scores you can use the function `rfit0` and for the rank dispersion criterion of the estimators the function `disp0`. Both functions can be found in Appendix A.

4.10 Consider the stackloss dataset from Example 4.4. Assuming that the variable `stack.loss` depends only on the variable `Acid.Conc.`, apply the methods used in Example 4.4 and compare the estimates with those in the example.

4.11 Software for robust statistics was developed by A. Marazzi (1992). The program library ROBETH has been interfaced to the statistical environments S-Plus and R (package *robeth*).

Chapter 5

Multivariate model

5.1 Concept of multivariate symmetry

The multivariate parametric estimation has been primarily elaborated for elliptically symmetric distributions, which appeared as an extension of the multi-normal distribution. However, not all symmetric multivariate distribution functions are elliptically symmetric. In the statistical practice and literature, we meet with other norms of the multivariate symmetry, each of them an interest and utilization of their own. Let us briefly describe the concepts of *diagonal symmetry, spherical symmetry, total symmetry, marginal symmetry,* and *symmetry in interchangeability.*

The basic tool in the multivariate model is the Mahalanobis distance. The squared Mahalanobis distance of $\mathbf{X}$ from $\boldsymbol{\theta}$ with respect to matrix $\boldsymbol{\Sigma}$ is

$$\|\mathbf{X} - \boldsymbol{\theta}\|_{\Sigma}^2 = (\mathbf{X} - \boldsymbol{\theta})^{\top} \boldsymbol{\Sigma}^{-1} (\mathbf{X} - \boldsymbol{\theta}) \tag{5.1}$$

Let $\mathbf{X}$ be a random vector taking values in $\mathbb{R}_p$, whose distribution function F has a density $f(\mathbf{x}; \boldsymbol{\theta}, \boldsymbol{\Sigma})$ which can be expressed as

$$f(\mathbf{x}; \boldsymbol{\theta}, \boldsymbol{\Sigma}) = c(\boldsymbol{\Sigma}) h_0(\|\mathbf{x} - \boldsymbol{\theta}\|_{\boldsymbol{\Sigma}}), \quad \mathbf{x} \in \mathbb{R}_p \tag{5.2}$$

where $h_0(y)$, $y \in \mathbb{R}^+$, is free from $\boldsymbol{\theta}, \boldsymbol{\Sigma}$. The *equi-probability contours* of density (5.2) are ellipsoidal with center $\boldsymbol{\theta}$ and orientation matrix $\boldsymbol{\Sigma}$. Such density is termed *elliptically symmetric.* If $h_0(\cdot)$ in (5.2) depends on $\mathbf{x}, boldgreek\theta$ through $\|\mathbf{x}-\boldsymbol{\theta}\|$ alone, we have a *spherically symmetric* density. The elliptically symmetric distributions, containing also the multivariate normal distribution, have finite second-order moments. However, not all symmetric multivariate distribution functions are elliptically symmetric.

We say that the distribution function $F(\mathbf{x}, \boldsymbol{\theta})$ is *diagonally symmetric,* if $\mathbf{X} - \boldsymbol{\theta}$ and $\boldsymbol{\theta} - \mathbf{X}$ have the same distribution function F_0, symmetric about $\mathbf{0}$. We also speak about *reflexive* or *antipodal* symmetry (Serfling 2010). It need not have the finite second moment or be elliptically symmetric; hence, $\mathcal{F}_E \subseteq \mathcal{F}_D$, where $\mathcal{F}_D$ denotes the family of all diagonal symmetric distributions and $\mathcal{F}_E$ the class of the elliptically symmetric distributions. In the subsequent text, we shall further take $\boldsymbol{\theta} = \mathbf{0}$, without loss of generality.

Let $\mathbf{r} = (r_1, \ldots, r_p)^\top$ be a vector with $0 - 1$ components. Denote $\mathcal{I}$ as the set of all possible $\mathbf{r}$'s and let $\mathbf{x} \circ \mathbf{r} = ((-1)^{r_1} x_1, \ldots, (-1)^{r_p} x_p)^\top$, $\mathbf{r} \in \mathcal{I}$. If

$$\mathbf{X} \circ \mathbf{r} \overset{\mathcal{D}}{=} \mathbf{X} \quad \forall \mathbf{r} \in \mathcal{I},$$

the distribution function F_0 of $\mathbf{X}$ is said to be *totally symmetric* about $\mathbf{0}$.

Another concept is the *marginal symmetry* of F : The distribution function F on $\mathbb{R}_p$ is called *marginally symmetric*, if all its marginal distribution functions $F_1, \ldots, F_p$ are symmetric about 0. The relation between the elliptical symmetry, diagonal symmetry, spherical symmetry and total symmetry can be written as

$$\mathcal{F}_E \subseteq \mathcal{F}_M \qquad \text{and} \quad \mathcal{F}_S \subseteq \mathcal{F}_T \subseteq \mathcal{F}_M \subseteq \mathcal{F}$$

where $\mathcal{F}_M$ is the class of marginally symmetric distributions.

Finally, let $\mathbf{x} \circ \mathbf{j} = (x_{j_1}, \ldots, x_{j_p})^\top$ where $\mathbf{j} = (j_1, \ldots, j_p)^\top$ is a permutation of $(1, \ldots, p)^\top$. We say that the coordinate variables of $\mathbf{X}$ are *exchangeable* or *interchangeable*, if $\mathbf{X} \circ \mathbf{j} \overset{\mathcal{D}}{=} \mathbf{X}$ for all permutations $\mathbf{j}$.

5.2   Multivariate location estimation

Let $\mathbf{X} \in \mathbb{R}_p$ be a random vector with a distribution function F. Consider an affine transformation $\mathbf{X} \mapsto \mathbf{Y} = \mathbf{A}\mathbf{X} + \mathbf{b}$ with $\mathbf{A}$ nonsingular of order $p \times p$, $\mathbf{b} \in \mathbb{R}_p$. Denote $F_{A,b}$ as the distribution function of $\mathbf{Y}$, also defined on $\mathbb{R}_p$. Select a vector-valued functional $\boldsymbol{\theta}(F)$ as a measure of the location of F. We call it an *affine-equivariant location functional*, provided

$$\boldsymbol{\theta}(F_{A,b}) = \mathbf{A}\boldsymbol{\theta}(F) + \mathbf{b} \quad \forall \mathbf{b} \in \mathbb{R}_p, \ \mathbf{A} \ \text{positive definite}.$$

Let $\boldsymbol{\Gamma}(F)$ be a matrix-valued functional of F, determined as a measure of the *scatter* of F around its location $\boldsymbol{\theta}$, describing its *shape* in terms of variation and covariance of the coordinate variables. We often call $\boldsymbol{\Gamma}(F)$ a *covariance functional*, and naturally assume that it is independent of $\boldsymbol{\theta}(F)$. We say that it is an *affine-equivariant covariance functional*, provided

$$\boldsymbol{\Gamma}(F_{A,b}) = \mathbf{A}\boldsymbol{\Gamma}(\mathbf{A})\mathbf{A}^\top \quad \forall \mathbf{b} \in \mathbb{R}_p, \ \mathbf{A} \ \text{positive definite}.$$

Whenever there exists a dispersion matrix $\boldsymbol{\Sigma}$, then $\boldsymbol{\Gamma}(F)$ should be proportional to $\boldsymbol{\Sigma}$, though the existence of the second moment may not be generally supposed.

Let $\mathbf{X}_1, \ldots, \mathbf{X}_n$ be independent observations of random vector $\mathbf{X}$. If F has a finite second moment, then the sample mean $\bar{\mathbf{X}} = (\bar{X}_{n1}, \ldots, \bar{X}_{np})^\top$, where $\bar{X}_{nj} = \frac{1}{n} \sum_{i=1}^{n} X_{ij}$, $j = 1, \ldots, p$, is an affine equivariant location functional, and the sample dispersion matrix

$$\mathbf{S}_n = n^{-1} \sum_{i=1}^{n} (\mathbf{X}_i - \bar{\mathbf{X}})(\mathbf{X}_i - \bar{\mathbf{X}})^\top$$

is an affine-equivariant covariance functional. However, $\bar{\mathbf{X}}$ and $\mathbf{S}_n$ are maximal likelihood estimators (MLE) of $\boldsymbol{\theta}$ and $\boldsymbol{\Sigma}$ only if F is multi-normal $(\boldsymbol{\theta}, \boldsymbol{\Sigma})$. In this situation $\bar{X}_{nj} = \frac{1}{n} \sum_{i=1}^{n} X_{ij}$ is also the marginal MLE of θ_j, $j = 1, \ldots, p$. If $f(\cdot, \cdot, cdot)$ in (5.2) is the multivariate Laplace density with $h_0(y) = \text{const} \cdot e^{-y}$, then the MLE of $\boldsymbol{\theta}$ is the spatial median, as we shall see in Section 8.10. Generally, the MLE of $\boldsymbol{\theta}$ heavily depends on the form of F and may be highly nonrobust to the model departures. This non-robustness may be further fortified by the presence of a multidimensional matrix nuisance parameter.

The location parameter $\boldsymbol{\theta}$ can be also estimated componentwise; we can apply the univariate L-, M- and R-estimation to each component of $\boldsymbol{\theta}$ separately. All the L-, M- and R- componentwise estimators have good robustness properties, L- and R-estimators being scale equivariant, but not the M-estimators. However, neither of them is affine-equivariant.

Let $F_1, \ldots, F_p$ denote the marginal distribution functions of F. Assume that $\theta_j = F_j^{-1}(\frac{1}{2})$ is uniquely defined for $j = 1, \ldots, p$; this happens if $F_j(x)$ is strictly monotone for x in a neighborhood of θ_j. Denote $\boldsymbol{\theta} = (\theta_1, \ldots, \theta_p)^\top$ as the vector of marginal medians. Let $X_{n1}, \ldots, X_{nj}$ be the n observations for the j-th coordinate and let $X_{j:1} < \ldots < X_{j:n}$ be the associated order statistics. Denote $\widehat{\boldsymbol{\theta}}_n = (\hat{\theta}_{n1}, \ldots, \hat{\theta}_{np})^\top$ as the vector of coordinate medians, where

$$\hat{\theta}_{nj} = \begin{cases} X_{j:\frac{n+1}{2}} & \text{for} \quad n \quad \text{odd} \\ \frac{1}{2}(X_{j:\frac{n}{2}} + X_{j:\frac{n}{2}+1}) & \text{for} \quad n \quad \text{even}, \ j = 1, \ldots, p. \end{cases} \tag{5.3}$$

Each $\hat{\theta}_{nj}$ is strongly consistent, which implies the strong consistency of $\widehat{\boldsymbol{\theta}}_n$. Because each $\hat{\theta}_{nj}$ is translation and scale equivariant, $\widehat{\boldsymbol{\theta}}_n$ is also translation equivariant, but not necessarily affine-equivariant.

The coordinatewise median $\widehat{\boldsymbol{\theta}}_n$ is a special case of coordinatewise M-, L- and R-estimators. The *coordinatewise R-estimators* are based on signed-rank statistics described in Section 3.10. Considering the multivariate analogues of M-, L- and R-estimators, we usually assume that the marginal distribution functions $F_1, \ldots, F_p$ are symmetric. Because the coordinates of $\mathbf{X}$ are mutually dependent, the breakdown points and other measures may not disseminate to the multivariate case.

5.3 Admissibility and shrinkage

Remember that an estimator $\widehat{\boldsymbol{\theta}}_n$ of $\boldsymbol{\theta}$ is admissible with respect to a specified loss function $L(\boldsymbol{\theta})$, if no other estimator has uniformly better or equal risk than $\widehat{\boldsymbol{\theta}}_n$, [i.e., $\mathrm{E}_\theta l(\boldsymbol{\theta})$] and is better at one point. In the univariate case, the sample mean $\bar{X}_n$ is an admissible estimator of the center of the normal distribution, and so even for more loss functions. However, the situation is different in the multivariate situation. Stein (1956) showed that under $p \geq 3$, the sample mean $\bar{\mathbf{X}}_n$ is not an admissible estimator of the center $\boldsymbol{\theta}$ of the normal distribution with respect to the quadratic loss function, even when it is the maximum likelihood estimator. Later on, James and Stein (1961) constructed an estimator that really dominates the MLE in the quadratic risk. In fact, there are infinitely many such estimators, called the shrinkage estimators, and they have been extensively studied. Estimators of multivariate location parameters are generally dominated by suitable shrinkage versions, and hence are inadmissible; even such shrinkage estimators themselves are not admissible.

Let us illustrate it in more detail. Stein (1981) studied the problem of estimating the mean of a multivariate normal distribution with the identity covariance matrix, based on single observation $\mathbf{X}$, and found a whole class of estimates dominating the naive estimator $\mathbf{X}$ with respect to the quadratic risk. A particularly important class of such estimators are James-Stein estimators $\left(1 - \frac{a}{\mathbf{X}^\top \mathbf{X}}\right) \mathbf{X}$, especially for $a = p - 2$ it has smaller quadratic risk than the naive estimator for all $\boldsymbol{\theta}$. In case of unknown scale, if there is an appropriate residual vector $\mathbf{U}$ estimating the scale, these estimators can be modified to $\left(1 - a \frac{\mathbf{U}^\top \mathbf{U}}{\mathbf{X}^\top \mathbf{X}}\right) \mathbf{X}$. As was shown by Cellier, Fourdrinier and Robert (1989) and by Cellier and Fourdrinier (1995), the latter estimates dominate $\mathbf{X}$ not only for the normal distribution, but with a proper a even simultaneously for all spherically symmetric distributions. Such estimates got the name "robust James-Stein estimators." They were later studied in more detail by Brandwein and Strawderman (1991), Fourdrinier and Strawderman (1996), Fourdrinier, Marchand and Strawderman (2004), and Kubokawa, Marchand and Strawderman (2014), among others. Jurečková and Milhaud (1993) derived a shrunken estimator for a class of totally symmetric densities, not necessary elliptically symmetric. This was further extended to the class of M-estimators by Jurečková and Sen (2006).

Mean square errors of the James-Stein estimator and the MLE for the multivariate normal distribution with the identity covariance matrix are compared in Table 5.1.

Table 5.1 *MSE estimates of James-Stein ($a = p - 2$ or 5) and ML estimates of the mean of p-variate normal distribution $\mathcal{N}_p(\boldsymbol{\theta}, \mathbf{I}_p)$ based on $100\,0000$ samples of size $n = 1$.*

	$\boldsymbol{\theta} = \mathbf{0}_p$			$\boldsymbol{\theta} = \mathbf{1}_p$		
Dimension	JS_{p-2}	ML	JS_5	JS_{p-2}	ML	JS_5
p=2	1.00	1.00	124.06	1.00	1.00	22.10
p=3	0.67	1.00	6.10	0.87	1.00	3.08
p=5	0.40	1.00	0.67	0.74	1.00	0.85
p=10	0.20	1.00	0.31	0.62	1.00	0.67
p=100	0.02	1.00	0.90	0.51	1.00	0.95

Let $\mathbf{X} = (X_1, \ldots, X_p)^\top$ be a random p-vector with an absolutely continuous distribution function $F(\mathbf{x}; \boldsymbol{\theta}) = F(x_1 - \theta_1, \ldots, x_p - \theta_p)$, $\mathbf{x} \in \mathbb{R}^p$, $\boldsymbol{\theta} \in \mathbb{R}^p$. Consider an estimator $\mathbf{T}_n$ of $\boldsymbol{\theta}$ based on a sample of n independent and identically distributed (i.i.d.) random vectors $\mathbf{X}_1, \ldots, \mathbf{X}_n$ drawn from F. The commonly used loss incurred due to estimating $\boldsymbol{\theta}$ by $\mathbf{T}_n$ is a general quadratic loss

$$L_Q(\mathbf{T}, \boldsymbol{\theta}) = (\mathbf{T}_n - \boldsymbol{\theta})^\top \mathbf{Q}^{-1}(\mathbf{T}_n - \boldsymbol{\theta}) \qquad (5.4)$$

with a positive definite matrix $\mathbf{Q}$; the corresponding risk is defined as

$$R_Q(\mathbf{T}, \boldsymbol{\theta}) = E L_Q(\mathbf{T}, \boldsymbol{\theta}) = \mathrm{Trace}(\mathbf{Q}^{-1}\mathbf{V}_T),$$

where

$$\mathbf{V}_T = \mathrm{E}(\mathbf{T} - \boldsymbol{\theta})(\mathbf{T} - \boldsymbol{\theta})^\top$$

is the dispersion matrix of $\mathbf{T}$. An important special case of the quadratic loss is the squared error loss corresponding to $\mathbf{Q} \equiv \mathbf{I}_p$,

$$L(\mathbf{T}, \boldsymbol{\theta}) = \|\mathbf{T} - \boldsymbol{\theta}\|^2 = \sum_{i=1}^n (T_i - \theta_i)^2.$$

If $\mathbf{Q} = \mathrm{diag}(q_{11}, \ldots, q_{pp})$ is a diagonal matrix, then (5.4) reduces to $\sum_{i=1}^p \frac{(T_i - \theta_i)^2}{q_{ii}}$. In practice, $\mathbf{Q}$ is our choice, while $\mathbf{V}_{T_n}$ depends on n and on the unknown F.

5.3.1   Reduction of risk of location estimator

Consider a random vector $\mathbf{X} = (X_1, \ldots, X_p)^\top$, $p \geq 3$, with density belonging to the class

$$f(x_1 - \theta_1, \ldots, x_p - \theta_p) = \prod_{i=1}^p f_i(x_i - \theta_i) = \exp\{-\sum_{i=1}^p \rho_i(x_i - \theta_i)\}, \qquad (5.5)$$

with $\boldsymbol{\theta} = (\theta_1, \ldots \theta_p)^\top$ being the true parameter, where ρ_i is an absolutely continuous convex function with derivative $\psi_i = \rho_i'$ and $\rho_i(-x) = \rho_i(x)$, $x \in \mathbb{R}_p$. We assume that

(a) either ψ_i has two bounded derivatives ψ_i', ψ_i'' in interval $(-c_i, c_i)$ and is constant outside $(-c_i, c_i)$, where $c_i > 0$ is a fixed number,

(b) or ψ_i has an absolutely continuous derivative ψ_i' in $\mathbb{R}_1$ and there exist $K > 0$ and $\delta > 0$ such that

$$\int_{\mathbf{R}^1} (\psi_i''(x+t))^2 e^{-\rho_i(x)} dx < K \quad \text{for } |t| < \delta.$$

Moreover, we assume that $0 < \mathcal{I}(f_i) = \int \psi_i^2(x) e^{-\rho_i(x)} dx < \infty$, $i = 1, \ldots, p$; but generally, the density in (5.5) need not be elliptically symmetric. We shall show that the L_2-risk of the random vector

$$\overline{\boldsymbol{\Psi}} = \left(\sum_{j=1}^n \psi_1(X_{1j} - \theta_1), \ldots, \sum_{j=1}^n \psi_p(X_{pj} - \theta_p) \right)^\top,$$

which is equal to

$$\mathbf{E}_{\boldsymbol{\theta}} \|(\sum_{j=1}^{n} \psi_1(X_{1j} - \theta_1), \ldots, \sum_{j=1}^{n} \psi_p(X_{pj} - \theta_p))^{\top}\|^2 = n \sum_{i=1}^{p} \mathcal{I}(f_i)$$

can be reduced by subtracting its posterior expectation, similar to that of the sample mean under the normal distribution. If X_i is normal $\mathcal{N}(\theta_i, \sigma_i^2)$, $i = 1, \ldots, p$, then

$$\overline{\boldsymbol{\Psi}}(\mathbf{X}, \boldsymbol{\theta}) = \left(\sigma_1^{-2} \sum_{j=1}^{n} (X_{1j} - \theta_1), \ldots, \sigma_p^{-2} \sum_{j=1}^{n} (X_{pj} - \theta_p)\right)^{\top}.$$

Under non-normal distribution,

$$\frac{1}{n}\overline{\boldsymbol{\Psi}}(\mathbf{X}, \mathbf{0}) = \frac{1}{n}\left(\sum_{j=1}^{n} \psi_1(X_{1j}), \ldots, \sum_{j=1}^{n} \psi_p(X_{pj})\right)^{\top}$$

cannot be considered as an estimator of $\boldsymbol{\theta}$; it is not linear and cannot be separated from unknown $\boldsymbol{\theta}$. However, when we derive its possible shrinkage, it will enable us to shrink the risk of some M-estimator!componentwises of location parameters asymptotically as $n \to \infty$; this will cover the maximal likelihood estimators.

We shall show that the L_2-norm of the random vector

$$\overline{\boldsymbol{\Psi}} = \left(\frac{1}{n} \sum_{j=1}^{n} \psi_1(X_{1j} - \theta_1), \ldots, \frac{1}{n} \sum_{j=1}^{n} \psi_p(X_{pj} - \theta_p)\right)^{\top} \qquad (5.6)$$

can be reduced under any fixed n, similar to that of the sample mean under the normal distribution. One possibility is to subtract from (5.6) its posterior expectation with respect to a superharmonic prior density.

The function g is called superharmonic if the value $g(\boldsymbol{\xi})$ at the center $\boldsymbol{\xi}$ of the ball $B(\boldsymbol{\xi}, \delta) \subset \mathbb{R}_p$ with radius δ is greater than or equal to the average of g over the boundary $\partial B(\boldsymbol{\xi}, \delta)$ of $B(\boldsymbol{\xi}, \delta)$, $\forall \boldsymbol{\xi}, \delta$. Denote $\nabla_i g = \frac{\partial g(\mathbf{x})}{\partial x_i}$, $\nabla_i^2 g = \frac{\partial^2 g(\mathbf{x})}{\partial x_i^2}$, $i = 1, \ldots, p$, $\nabla g = (\nabla_1 g, \ldots, \nabla_p g)^{\top}$, and $\Delta g = \sum_{i=1}^{p} \nabla_i^2 g$ (whenever it exists). If g is superharmonic, then $\Delta g(\mathbf{x}) \le 0$ $\forall \mathbf{x} \in \mathbb{R}_p$ (see Helms (1969), Theorem 4.8).

Consider (5.5) as the conditional density of $\mathbf{X}$ given the parameter value $\boldsymbol{\theta}$ and let Π be a prior probability distribution on the Borel subsets of $\mathbb{R}_p$ with a pseudoharmonic Lebesgue density π. Then the unconditional density of $\mathbf{X}$ with respect to Π is

$$h(\mathbf{x}) = \int_{\mathbb{R}_p} \exp\left\{-\sum_{i=1}^{p} \rho_i(x_i - z_i)\right\} \pi(\mathbf{z}) d\mathbf{z} = \int_{\mathbb{R}_p} \exp\left\{-\sum_{i=1}^{p} \rho_i(y_i)\right\} \pi(\mathbf{x} - \mathbf{y}) d\mathbf{y}.$$

$$(5.7)$$

Assume that

(P.1) $\pi(z_1, \ldots, z_p)$ is a twice continuously differentiable superharmonic density.

(P.2) $\lim_{\|\mathbf{z}\| \to \infty} \pi(z_1, \ldots, z_p) = 0$.

The existence of π satisfying **(P.1)** and **(P.2)** along with a detailed exposition of superharmonic functions can be found in Blumental and Getoor (1968), Helms (1969), Chung (1982), and Doob (1984), among others. Remember that π satisfying **(P.1)** and **(P.2)** is superharmonic if and only if

$$\Delta \pi \leq 0, \quad \text{where} \quad \Delta \pi = \sum_{i=1}^{p} \frac{\partial^2 \pi(\mathbf{z})}{\partial z_i^2} \quad \text{is the Laplacian of } \pi. \quad (5.8)$$

Besides superharmonic functions cited in the above literature, Jurečková and Sen (2006) proposed another proper superharmonic prior based on a mixture of normal densities.

Let $\mathbf{X}_1, \ldots, \mathbf{X}_n$ be a random sample from distribution (5.5). Then the simultaneous density of $\mathbb{X} = (\mathbf{X}_1, \ldots, \mathbf{X}_n)^\top$ is

$$g(\mathbb{X}, \boldsymbol{\theta}) = \exp\{-\sum_{j=1}^{n} \sum_{i=1}^{p} \rho_i(x_{ij} - \theta_i)\} \quad (5.9)$$

Considering again the density (5.9) as a conditional density of $\mathbb{X} = (\mathbf{X}_1, \ldots, \mathbf{X}_n)^\top$ given $\boldsymbol{\theta}$, we get its marginal density (with respect to Π) in the form

$$g^*(\mathbb{X}) = \int_{\mathbb{R}_p} \exp\left\{ -\sum_{j=1}^{n} \sum_{i=1}^{p} \rho_i(x_{ij} - z_i) \right\} \pi(\mathbf{z}) d\mathbf{z}. \quad (5.10)$$

The conditional or posterior density of $\boldsymbol{\theta}$, given $\mathbb{X}$, is

$$\frac{1}{g^*(\mathbb{X})} \pi(\boldsymbol{\theta}) \exp\left\{ -\sum_{i=1}^{n} \sum_{j=1}^{p} \rho_i(x_{ij} - \theta) d\theta) \right\}.$$

As a result, the posterior expectation of $\sum_{j=1}^{p} \psi_i(X_{ij} - \theta_i)$, given $\mathbb{X}$, is equal to

$$\sum_{j=1}^{n} \frac{\partial \log g^*(\mathbb{X})}{\partial X_{ij}} = -\mathrm{E}^{\mathbf{X}}\left(\sum_{j=1}^{n} \psi_i(X_{ij} - Z_i) \right), \quad i = 1, \ldots, p \quad (5.11)$$

where $\mathrm{E}^{\mathbf{X}}(\cdot)$ denotes the posterior expectation with respect to prior Π. It is equal to

$$-\frac{1}{g^*(\mathbb{X})} \int_{\mathbb{R}^p} \sum_{j=1}^{n} \psi_i(X_{ij} - z_i) \exp\left\{ -\sum_{k=1}^{p} \sum_{\nu=1}^{n} \rho_k(X_{k\nu} - z_k) \right\} \pi(\mathbf{z}) d\mathbf{z}. \quad (5.12)$$

The following theorem shows that the quadratic risk of the random vector

$$\sum_{j=1}^{n} (\psi_1(X_{1j} - \theta_1), \ldots, \psi_p(X_{pj} - \theta_p))^\top$$

under $\boldsymbol{\theta}$ can be reduced by subtracting its posterior expectation, similar to the multivariate normal sample mean problem.

Theorem 5.1 *Assume that the density (5.5) satisfies (**a**) and (**b**) and that the prior density π satisfies (**P.1**) and (**P.2**). Then, defining the posterior expectation as in (5.12),*

$$
\mathrm{E}_{\boldsymbol{\theta}}\Big\{ \sum_{i=1}^{p} \Big[\sum_{j=1}^{n} \psi_i(X_{ij} - \theta_i) - \mathrm{E}^{\mathbb{X}}\Big(\sum_{j=1}^{n} \psi_i(X_{ij} - Z_i) \Big) \Big]^2 \Big\}
$$

(5.13)

$$
\leq \mathrm{E}_{\boldsymbol{\theta}}\Big\{ \sum_{i=1}^{p} \Big[\sum_{j=1}^{n} \psi_i(X_{ij} - \theta_i) \Big]^2 \Big\}.
$$

For the proof we refer to Jurečková and Sen (2006).

This result yields an asymptotic shrinkage of some estimators under $n \to \infty$ in a neighborhood of a suitable pivot. The following lemma shows an application to the maximum likelihood estimation of $\boldsymbol{\theta}$:

Lemma 5.1 *Let $\mathbb{X} = (\mathbf{X}_1, \ldots, \mathbf{X}_n)^\top$ be a random sample from the distribution with density (5.5) satisfying (**a**) and (**b**) with $\rho_1 \equiv \rho_2 \equiv \ldots \equiv \rho_p$. Let π be the prior density satisfying (**P.1**) and (**P.2**). Then, for the MLE $\mathbf{T}_n$ of $\boldsymbol{\theta}$,*

$$
\sum_{i=1}^{p} \mathrm{E}_{\boldsymbol{\theta}}\Big\{ T_{ni} - \theta_i - \mathrm{E}^{\mathbb{X}}\Big(\frac{1}{n\mathcal{I}(f)} \sum_{j=1}^{n} \psi(X_{ij} - Z_i) \Big) \Big\}^2 = \frac{p}{n\mathcal{I}(f)} - \frac{K_\theta^2}{n^2} + o(n^{-2})
$$

as $n \to \infty$,

where

$$
K_\theta = \sum_{i=1}^{p} \mathrm{E}_{\boldsymbol{\theta}}\Big\{ \sum_{j=1}^{n} \frac{1}{(g^*(\mathbb{X}))^2} \Big(\int_{\mathbb{R}^p} g(\mathbb{X}, \mathbf{z}) \nabla_i \pi(\mathbf{z}) d\mathbf{z} \Big)^2 \Big\}
$$

$$
- 2\mathrm{E}_{\boldsymbol{\theta}}\Big[\frac{1}{g^*(\mathbb{X})} \int_{\mathbb{R}^p} g(\mathbb{X}, \mathbf{z}) \Delta \pi(\mathbf{z}) d\mathbf{z} \Big].
$$

(5.14)

The lemma is proven in Jurečková and Milhaud (1993) in detail.

5.4   Visualization of multivariate data in R

Software R provides very good tools for all kinds of graphics. There are many packages designed for the graphical display of multivariate data of different types. Here we mention only some of the packages. The base package called *graphics* includes useful functions like `pairs()` for a scatterplot matrix, `coplot()` for a conditioning plot or `mosaicplot()` for categorical data.

There are also packages like *lattice* or *ggplot2* that are based on grid graphics which is a more advanced system than the standard R graphics. The *lattice* package can, with simple code, convert multivariate data into complex

graphics, see the book by Sarkar (2008). In particular, it can produce very nice
multi-panel plots displaying the relationship between variables conditioned on
other variables. The *ggplot2* is another very popular package for creating beau-
tiful and complex graphics, see Wickham (2009). A plot is created iteratively
by adding separate layers with a + sign.

We use the dataset SAT from the package *mosaicData* for illustration. The
SAT is a standardized college admission test in the United States. For each
of the 50 U.S. states the data includes the average total SAT score (*sat*),
expenditure per pupil (*expend*), average pupil/teacher ratio (*ratio*), average
salary of teachers (*salary*), and the percentage of all eligible students taking
the SAT (*frac*). The remaining three variables (the first, the sixth and the
seventh column) are not displayed. The following code was used to create
Figure 5.1.

```
> library(mosaicData)
> pairs(SAT[,-c(1,6:7)])
```

The scatterplot matrix consists only of panels of bivariate relationship.
From the lower left panel of Figure 5.1 it seems that the average SAT score is
lower in high-spending states than in the low-spending states, on average.

In this case it is useful to produce a plot conditioned on the variable
frac. We can use a function xyplot() from the lattice package to produce
Figure 5.2.

```
> library(lattice)
> library(quantreg)
> Frac <- equal.count(SAT$frac, number=4, overlap=0)
> xyplot(sat ~ expend | Frac, data = SAT,
+          panel = function(x, y)
+                    { panel.xyplot(x, y)
+                      panel.abline(rq(y~x))},
+          strip = strip.custom(strip.names = TRUE,
+                                 strip.levels = TRUE),
+          par.strip.text = list(cex = 0.85))
```

In Figure 5.2 the range of *frac* is divided into four (possibly overlapping)
intervals of approximately equal size. For this a new vector *Frac* is created by
the function equal.size. In the first argument of the function we specify that
we want to plot *sat* versus *expend* conditional on the variable *Frac*. Each panel
then contains observations from only one interval specified in *Frac*. Moreover,
regression median lines are drawn. For that, the *panel* function that does the
actual plotting must be added. Inside the function we specify that on each
panel we want to plot the data (*panel.xyplot*) and a regression quantile line

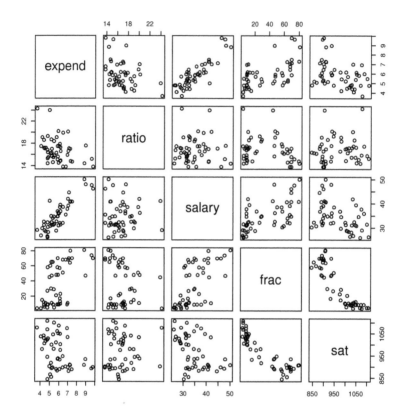

Figure 5.1 *Scatterplot matrix of multivariate data.*

(*panel.abline*) computed by *rq* from a package *quantreg*. This gives us better understanding of what is the true dependence of *sat* on *expend*. We use also the argument *strip* to draw the name as well as the levels of the variable *Frac* and *par.strip.text* to make the text smaller. Obviously, this kind of plot can reveal relationships that would be impossible to see from the traditional scatterplot matrix.

In Figure 5.3 we present a similar plot, here including least square lines and their confidence intervals, by ggplot() from the *ggplot2* package. The code to generate the figure is the following.

```
> library(ggplot2)
> FRAC <- cut(SAT$frac, breaks = quantile(SAT$frac),
+             include.lowest = T)
> ggplot(SAT, aes(expend, sat) ) +
+   geom_point( aes(shape = FRAC) , size = 2) +
```

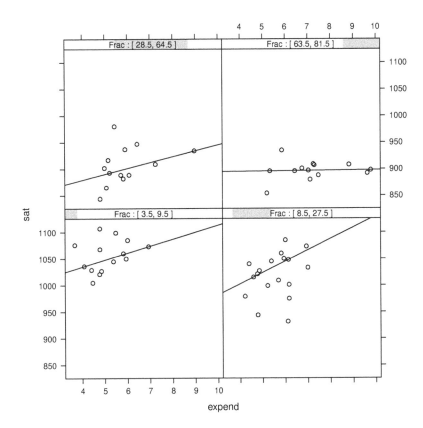

Figure 5.2 *Example plot using lattice package.*

```
+    geom_smooth( aes(linetype = FRAC), method = "lm") +
+    scale_shape(solid = FALSE) + ggtitle('SAT scores') +
+    theme(plot.title = element_text(hjust = 0.5))
```

First a vector *FRAC* is created with a function *cut* that divides the range of *frac* into four groups with quartiles as endpoints. With *ggplot*, the plot object is initialized. There we also specify the variables that are common for the following layers. The layer *geom_point* draws the scatterplot and the shape of points is determined by the level of *FRAC*. Next *geom_smooth* is used to create the least square lines (different for each level of *FRAC*) with the confidence intervals, and with the layer *scale_shape*, the shapes are changed to be hollow. The last two layers are added to produce a centered main title of the plot.

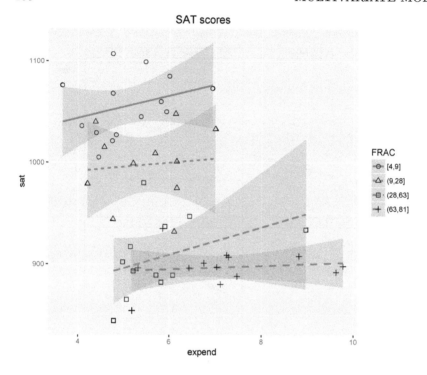

Figure 5.3 *Example plot using ggplot2 package.*

For 3d plots there are packages like *scatterplot3d*, or a very powerful package *rgl*.

5.5 Multivariate regression estimation

Consider independent p-vectors $\mathbf{Y}_1, \ldots, \mathbf{Y}_n$ and denote $\mathbb{Y} = [\mathbf{Y}_1, \ldots, \mathbf{Y}_n]$ as the $p \times n$ random matrix. Assume that the $\mathbf{Y}_i = (Y_{i1}, \ldots, Y_{ip})^\top$ depends on a non-stochastic regressor $\mathbf{x}_i \in \mathbb{R}_q$ with known components, not all equal, in the form $\mathbf{Y}_i = \boldsymbol{\beta} \mathbf{x}_i + \mathbf{e}_i$, $1 \leq i \leq n$. Here $\boldsymbol{\beta}$ is a $p \times q$ matrix of unknown regression parameters and $\mathbf{e}_1, \ldots, \mathbf{e}_n$ are i.i.d. random vectors. Denote $\mathbb{X} = [\mathbf{x}_1, \ldots, \mathbf{x}_n]$ as the matrix of order $q \times n$ and write the regression model as

$$\mathbb{Y} = \boldsymbol{\beta} \mathbb{X} + \mathbb{E}; \quad \mathbb{E} = [\mathbf{e}_1, \ldots, \mathbf{e}_n] \tag{5.15}$$

The parameter $\boldsymbol{\beta}$ is often partitioned as $(\boldsymbol{\theta}, \boldsymbol{\beta}^*)$ with $\boldsymbol{\beta}^*$ of order $p \times (q-1)$; then the top row of $\mathbb{X}$ is $\mathbf{1}_n^\top = (1, \ldots, 1)$, and $\boldsymbol{\theta}$ is regarded as the vector of intercept parameters, while $\boldsymbol{\beta}^*$ is the matrix of regression parameters. If $q = 1$, then (5.15) relates to the multivariate location model.

5.5.1 Multivariate linear model

If $\mathbf{e}_1, \ldots, \mathbf{e}_n$ have a multi-normal distribution with null mean vector and a positive definite dispersion matrix $\boldsymbol{\Sigma}$ and

$$\|\mathbf{Y}_i - \boldsymbol{\beta}\mathbf{x}_i\|_{\boldsymbol{\Sigma}}^2 = (\mathbf{Y}_i - \boldsymbol{\beta}\mathbf{x}_i)^\top \boldsymbol{\Sigma}^{-1}(\mathbf{Y}_i - \boldsymbol{\beta}\mathbf{x}_i), \quad i = 1, \ldots, n,$$

then the MLE of $\boldsymbol{\beta}$ for the normal model is the minimizer of $\sum_{i=1}^n \|\mathbf{Y}_i - \boldsymbol{\beta}\mathbf{x}_i\|_{\boldsymbol{\Sigma}}^2$ and is given by

$$\widehat{\boldsymbol{\beta}}_n = (\mathbb{Y}\mathbb{X}^\top)(\mathbb{X}\mathbb{X}^\top)^{-1}.$$

$\widehat{\boldsymbol{\beta}}_n$ is a linear estimate of $\boldsymbol{\beta}$ and is unbiased and affine-equivariant both in $\mathbb{Y}$ and $\mathbb{X}$.

Even if $\mathbf{e}_1, \ldots, \mathbf{e}_n$ are not multinormal but i.i.d. and $\mathrm{E}(\mathbf{e}_i) = \mathbf{0}$, $\mathrm{E}(\mathbf{e}_i\mathbf{e}_i^\top) = \boldsymbol{\Sigma}$ (positive definite), and

$$\max_{1 \le i \le n} \mathbf{x}^\top (\mathbb{X}\mathbb{X}^\top)^{-1}\mathbf{x}_i \to 0 \quad \text{as} \quad n \to \infty$$

(the generalized Noether condition) is satisfied, $\widehat{\boldsymbol{\beta}}_n$ is a linear estimator of $\boldsymbol{\beta}$, (double)-affine equivariant and is asymptotically normally distributed. However, it is highly nonrobust.

As an alternative, we can consider the coordinatewise robust estimators in the multivariate linear regression model. The finite-sample as well as the asymptotic properties of regression L-, M- and R-estimators pass componentwise to the multivariate case.

In Section 5.6 we shall propose the robust affine equivariant L-estimator in the multivariate linear regression model. The loss based on the L_2-norm will be replaced by the L_1-norm and as the estimator we take the minimizer of $\sum_{i=1}^n \|\mathbf{Y}_i - \boldsymbol{\beta}\mathbf{x}_i\|_{\boldsymbol{\Sigma}}$.

5.6 Affine invariance and equivariance, maximal invariants

Let $\mathbb{X}_n = (\mathbf{X}_1, \ldots, \mathbf{X}_n)$ be a random sample from p-variate population with continuous distribution function F with location parameter $\boldsymbol{\theta}$ and dispersion matrix $\boldsymbol{\Sigma}$. Consider the set of affine transformations

$$\mathcal{G} : \{\mathbf{X} \to \mathbf{a} + \mathbf{B}\mathbf{X}\} \quad \text{with } \mathbf{a} \in \mathbb{R}_p \quad \text{and nonsingular } p \times p \text{ matrix } \mathbf{B}. \quad (5.16)$$

Obenchain (1971) showed that the maximal invariant with respect to $\mathcal{G}$ is

$$\mathbf{S}_n(\mathbf{X}_1, \ldots, \mathbf{X}_n) = \left[(\mathbf{X}_i - \bar{\mathbf{X}}_n)^\top \mathbf{V}_n^{-1}(\mathbf{X}_j - \bar{\mathbf{X}}_n)\right]_{i,j=1}^n, \quad (5.17)$$

$$\text{where} \quad \bar{\mathbf{X}}_n = \frac{1}{n}\sum_{i=1}^n \mathbf{X}_i, \ \mathbf{V}_n = \sum_{i=1}^n (\mathbf{X}_i - \bar{\mathbf{X}}_n)(\mathbf{X}_i - \bar{\mathbf{X}}_n)^\top.$$

Then $\mathbf{S}_n(\mathbf{X}_1, \ldots, \mathbf{X}_n)$ is the projection matrix associated with the space spanned by the columns of the matrix $\left[\mathbf{X}_1 - \bar{\mathbf{X}}_n, \ldots, \mathbf{X}_n - \bar{\mathbf{X}}_n\right]$. Especially, under $\mathbf{a} \equiv \mathbf{0}$, the maximal invariant of the group of transformations

$$\mathcal{G}_0 : \{\mathbf{X} \to \mathbf{B}\mathbf{X}\} \tag{5.18}$$

is equal to $\mathbf{T}_0(\mathbf{X}_1, \dots, \mathbf{X}_n) = \left[\mathbf{X}_i^{\top}(\mathbf{V}_n^0)^{-1}\mathbf{X}_j\right]_{i,j=1}^{n}$, where $\mathbf{V}_n^0 = \sum_{i=1}^{n} \mathbf{X}_i \mathbf{X}_i^{\top}$.

Moreover, one of the maximal invariants with respect to the *group of shifts in location* $\mathcal{G}_1$: $\mathbf{X} \longrightarrow \mathbf{X} + \mathbf{a}$, $\mathbf{a} \in \mathbb{R}_p$ is $\mathbf{T}_1(\mathbf{X}_1, \dots, \mathbf{X}_n) = (\mathbf{X}_2 - \mathbf{X}_1, \dots, \mathbf{X}_n - \mathbf{X}_1)$.

Hence, every statistic $\mathbf{T}(\mathbf{X}_1, \dots, \mathbf{X}_n)$ affine-invariant with respect to $\mathcal{G}$ depends on $\mathbf{X}_1, \dots, \mathbf{X}_n$ only by means of $\mathbf{S}_n$, thus it is a function of $\mathbf{S}_n$. The Mahalanobis distance (5.1) plays a central role in the development of the affine equivariance. It is a natural follower of the Euclidean norm $\|\mathbf{X} - \boldsymbol{\theta}\|$, which is rotation invariant but not affine-invariant. However, the Mahalanobis distance cannot be shortened by using a robust $\hat{\boldsymbol{\theta}}_n$ instead of $\overline{\mathbf{X}}_n$.

The robust estimators are nonlinear, hence not automatically affine equivariant; for that they would need a special approach. A novel methodology has been developed for the spatial median, spatial rank and spatial quantile functions; we refer to Oja (2010) and Serfling (2010), and to other references cited therein. We also refer to Liu at al. (1999), Chakraborty (2001), Hallin et al. (2010), Kong and Mizera (2012) and others for various approaches to multivariate quantiles. The L_1-estimators were considered by Roelant and van Aelst (2007) and by Sen et al. (2013). Robust affine-equivariant estimators with a high breakdown point were studied by Lopuhaä and Rousseeuw (1991) and by Zuo (2003, 2006), among others.

5.6.1 *Multivariate sign and ranks*

Let $\mathcal{B}_{p-1}(\mathbf{0})$ be the open unit ball and let $\Phi(\mathbf{u}, \mathbf{y}) = \|\mathbf{u}\| + \langle \mathbf{u}, \mathbf{y} \rangle$, $\mathbf{u} \in \mathcal{B}_{p-1}(\mathbf{0})$, $\mathbf{y} \in \mathbb{R}_p$. The $\mathbf{u}$-th *spatial quantile* $Q_F(\mathbf{u})$ is defined as the minimizer of the objective function

$$\bar{\Psi}(\mathbf{u}) = \mathrm{E}\{\Phi(\mathbf{u}, \mathbf{X} - \boldsymbol{\theta}) - \Phi(\mathbf{u}, \mathbf{X}), \ \mathbf{X} \in \mathbb{R}_p, \ \mathbf{u} \in \mathcal{B}_{p-1}$$

in $\boldsymbol{\theta}$. More precisely, it is a solution $Q_F(\mathbf{u}) = \boldsymbol{\xi}_u$ of the equation

$$\mathbf{u} = \mathrm{E}\left\{\frac{\boldsymbol{\xi}_u - \mathbf{X}}{\|\boldsymbol{\xi}_u - \mathbf{X}\|}\right\}.$$

Particularly, $Q_F(\mathbf{0})$ is the spatial median. To any $\mathbf{x} \in \mathbb{R}_p$ there exists a unit vector $\mathbf{u_x}$ such that $\mathbf{x} = Q_F(\mathbf{u_x})$. Hence,

$$\mathbf{u_x} = Q_F^{-1}(\mathbf{x}) \text{ is the } inverse \ spatial \ quantile \ functional \text{ at } \mathbf{x}.$$

Note that $\mathbf{u}_\theta = \mathbf{0}$ at $\boldsymbol{\xi} = \boldsymbol{\theta}$, so that $Q_F(\mathbf{0}) = \boldsymbol{\theta}$. The quantile $Q_F(\mathbf{u})$ is central for $\mathbf{u} \sim \mathbf{0}$ and it is extreme for $\mathbf{u} \sim \mathbf{1}$. The values $Q_F^{-1}(\mathbf{x})$ can coincide for multiple $\mathbf{x}$; moreover, Möttönen and Oja (1995) showed that $Q_F(\cdot)$ and $Q_F^{-1}(\cdot)$ are inverse to each other.

The sample counterpart of the population spatial quantile function is defined as a solution $\widehat{\mathbf{x}}_{n,u}$ of the equation

$$\frac{1}{n} \sum_{i=1}^{n} \frac{\mathbf{x} - \mathbf{X}_i}{\|\mathbf{x} - \mathbf{X}_i\|} = \mathbf{u}, \quad \text{i.e., } \widehat{\mathbf{x}}_{n,u} = Q_n^{-1}(\mathbf{u})$$

where $\frac{\mathbf{a}}{\|\mathbf{a}\|}$ is defined as $\mathbf{0}$ if $\mathbf{a} = \mathbf{0}$, similar to the sign function. The function

$$R_n^*(\mathbf{x}) = \frac{1}{2n} \sum_{i=1}^{n} \left[\frac{\mathbf{x} - \mathbf{X}_i}{\|\mathbf{x} - \mathbf{X}_i\|} + \frac{\mathbf{x} + \mathbf{X}_i}{\|\mathbf{x} + \mathbf{X}_i\|} \right], \quad \mathbf{x} \in \mathbb{R}_p$$

is defined as the *sample central rank function*. The spatial median is equivariant with respect to the shift and to the orthogonal and homogeneous scale transformations. Indeed, let $\mathbf{y} = \mathbf{Bx} + \mathbf{a}b$, $\mathbf{a} \in \mathbb{R}_p$, $\mathbf{B}$ positive definite orthogonal. Then

$$\mathbf{u} \mapsto \mathbf{u}' = \frac{\|\mathbf{u}\|}{\|\mathbf{Bu}\|} \cdot \mathbf{Bu}, \quad \mathbf{u} \in \mathcal{B}_{p-1}(\mathbf{0})$$

and if $F_{B,a}$ is the distribution function of $\mathbf{Y}$, then $Q_{F_{B,a}}(\mathbf{u}') = \mathbf{B}Q_F(\mathbf{u}) + \mathbf{b}$; also if $\mathbf{u} = \mathbf{0}$ then $\mathbf{u}' = \mathbf{0}$. However, the spatial quantile function is generally not affine-equivariant for all $\mathbf{u}$.

5.6.2 *Affine-equivariant robust estimators*

Let us denote

$$d_{ni} = (\mathbf{X}_i - \overline{\mathbf{X}}_n)^\top \mathbf{V}_n^{-1}(\mathbf{X}_i - \overline{\mathbf{X}}_n), \ i = 1, \ldots, n \qquad (5.19)$$

as the diagonal elements of matrix (5.26), where $\mathbf{V}_n = \sum_{i=1}^{n}(\mathbf{X}_i - \overline{\mathbf{X}}_n)(\mathbf{X}_i - \overline{\mathbf{X}}_n)^\top$. All the d_{ni} are between 0 and 1 and their sum equals p. Because d_{ni} are exchangeable, bounded random variables, all nonnegative, with a constant sum equal to p, the asymptotic properties of the array $(d_{n1}, \ldots, d_{nn})^\top$ follow from Chernoff and Teicher (1958) and Weber (1980). Let R_{ni} be the rank of d_{ni} among $d_{n1}, \ldots, d_{nn}$, $i = 1, \ldots, n$, i.e., $R_{ni} = \sum_{j=1}^{n} I[d_{nj} \le d_{ni}]$. Denote $\mathbf{R}_n = (R_{n1}, \ldots, R_{nn})^\top$ as the vector of ranks. The probability of ties is 0, because F is continuous, hence the ranks are well defined. All d_{ni} and R_{ni} are affine-invariant, $i = 1, \ldots, n$. Moreover, the R_{ni} are invariant under any strictly monotone transformation of d_{ni}, $i = 1, \ldots, n$. Each $\mathbf{X}_i$ is trivially affine-equivariant. We introduce the following (Mahalanobis) ordering of $\mathbf{X}_1, \ldots, \mathbf{X}_n$:

$$\mathbf{X}_i \prec \mathbf{X}_j \Leftrightarrow d_{ni} < d_{nj}, \ i \ne j = 1, \ldots, n \qquad (5.20)$$

which is also affine invariant. Then we can consider vector of order statistics $\mathbf{X}_{n:1} \prec \ldots \prec \mathbf{X}_{n:n}$ of the sample $\mathbb{X}_n$, based on this ordering. The smallest d_{ni} corresponds to *the nearest neighborhood of the center*. Take a sequence $\{k_n\}$

of nonnegative integers such that k_n is $\nearrow$ in n, but $n^{-1/2}k_n$ is $\searrow$ in n, and for fixed k put

$$\mathbf{L}_{nk} = \binom{k_n}{k}^{-1} \sum_{i=1}^{n} I\left[R_{ni} \le k_n\right] \binom{k_n - R_{ni}}{k-1} \mathbf{X}_i.$$

$\mathbf{L}_{nk}$ is affine-equivariant, because the d_{ni} are affine invariant and the $\mathbf{X}_i$ are trivially affine equivariant. Particularly, for $k = 1$,

$$\mathbf{L}_{n1} = k_n^{-1} \sum_{i=1}^{n} I\left[R_{ni} \le k_n\right] \mathbf{X}_i$$

represents a trimmed, rank-weighted, nearest neighbor (NN) affine-equivariant estimator of $\boldsymbol{\theta}$. In the case $k = 2$ we have

$$\mathbf{L}_{n2} = \binom{k_n}{2}^{-1} \sum_{i=1}^{n} I\left[R_{ni} \le k_n\right] (k_n - R_{ni})\mathbf{X}_i$$

which can be rewritten as $\quad \mathbf{L}_{n2} = \sum_{i=1}^{n} w_{nR_{ni}}\mathbf{X}_i$ with the weight-function

$$w_{ni} = \begin{cases} \binom{k_n}{2}^{-1} (k_n - i) & \dots \quad i = 1, \dots, k_n; \\ 0 & \dots \quad i > k_n. \end{cases}$$

A greater influence is put on for $R_{ni} = 1$ or 2, and $w_{nk_n} = 0$; $w_{n1} = 2/k_n$; this is also for $k \ge 3$. The weights $w_n(i)$ can be chosen as the nonincreasing rank scores $a_n(1) \ge a_n(2) \ge \dots \ge a_n(n)$, e.g., the Wilcoxon ones.

To construct an affine equivariant L-estimator of the location parameter, we start with initial affine-equivariant location estimator and scale functional, and then iterate it to a higher robustness. For simplicity we start with the sample mean vector $\overline{\mathbf{X}}_n = \frac{1}{n}\sum_{i=1}^{n} \mathbf{X}_i$ and with the matrix $\boldsymbol{V}_n = \mathbf{A}_n^{(0)} = \sum_{i=1}^{n}(\mathbf{X}_i - \overline{\mathbf{X}}_n)(\mathbf{X}_i - \overline{\mathbf{X}}_n)^\top = n\widehat{\boldsymbol{\Sigma}}_n$, $n > p$, and continue recursively: Put $\mathbf{L}_n^{(1)} = \mathbf{L}_n$ and define in the next step:

$$\mathbf{A}_n^{(1)} = \sum_{i=1}^{n}(\mathbf{X}_i - \mathbf{L}_n^{(1)})(\mathbf{X}_i - \mathbf{L}_n^{(1)})^\top$$

$$d_{ni}^{(1)} = (\mathbf{X}_i - \mathbf{L}_n^{(1)})^\top (\mathbf{A}_n^{(1)})^{-1}(\mathbf{X}_i - \mathbf{L}_n^{(1)})$$

$$R_{ni}^{(1)} = \sum_{j=1}^{n} I[d_{nj}^{(1)} \le d_{ni}^{(1)}], \ i = 1, \dots, n, \quad \mathbf{R}_n^{(1)} = (R_{n1}^{(1)}, \dots, R_{nn}^{(1)})^\top.$$

The second-step estimator is $\mathbf{L}_n^{(2)} = \sum_{i=1}^{n} w_n(R_{ni}^{(1)})\mathbf{X}_i$. In this way we proceed, so at the r-th step we define $\mathbf{A}_n^{(r)}$, $d_{ni}^{(r)}$, $1 \le i \le n$ and the ranks $\mathbf{R}_n^{(r)}$ analogously, and get the r-step estimator

$$\mathbf{L}_n^{(r)} = \sum_{i=1}^{n} w_n(R_{ni}^{(r-1)})\mathbf{X}_i, \ r \ge 1. \tag{5.21}$$

The algorithm proceeds as follows:

(1) Calculate $\overline{\mathbf{X}}_n$ and $\mathbf{A}_n^{(0)} = \sum_{i=1}^n (\mathbf{X}_i - \overline{\mathbf{X}}_n)(\mathbf{X}_i - \overline{\mathbf{X}}_n)^\top$.

(2) Calculate $d_{ni}^{(0)} = (\mathbf{X}_i - \overline{\mathbf{X}}_n)^\top (\mathbf{A}_n^{(0)})^{-1}(\mathbf{X}_i - \overline{\mathbf{X}}_n)$, $1 \le i \le n$.

(3) Determine the rank $R_{ni}^{(0)}$ of $d_{ni}^{(0)}$ among $d_{n1}^{(0)}, \ldots, d_{nn}^{(0)}$, $i = 1, \ldots, n$.

(4) Calculate the scores $a_n(i)$, $i = 1, \ldots, n$

(5) Calculate the first-step estimator $\mathbf{L}_n^{(1)} = \sum_{i=1}^n a_n(R_{ni}^{(0)})\mathbf{X}_i$.

(6) $\mathbf{A}_n^{(1)} = \sum_{i=1}^n (\mathbf{X}_i - \mathbf{L}_n^{(1)})(\mathbf{X}_i - \mathbf{L}_n^{(1)})^\top$.

(7) $d_{ni}^{(1)} = (\mathbf{X}_i - \mathbf{L}_n^{(1)})^\top (\mathbf{A}_n^{(1)})^{-1}(\mathbf{X}_i - \mathbf{L}_n^{(1)})$, $1 \le i \le n$.

(8) $R_{ni}^{(1)} =$ the rank of $d_{ni}^{(1)}$ among $d_{n1}^{(1)}, \ldots, d_{nn}^{(1)}$, $i = 1, \ldots, n$.

(9) $\mathbf{L}_n^{(2)} = \sum_{i=1}^n a_n(R_{ni}^{(1)})\mathbf{X}_i$.

(10) Repeat the steps (6)–(9).

The estimator $\mathbf{L}_n^{(r)}$ is a linear combination of order statistics corresponding to independent random vectors $\mathbf{X}_1, \ldots, \mathbf{X}_n$, with random coefficients based on the exchangeable $d_{ni}^{(r)}$. The estimating procedure preserves the affine equivariance at each step and $\mathbf{L}_n^{(r)}$ is an affine-equivariant L-estimator of $\boldsymbol{\theta}$ for every r. Indeed, if $\mathbf{Y}_i = \mathbf{B}\mathbf{X}_i + \mathbf{a}$, $\mathbf{a} \in \mathbb{R}_p$, $\mathbf{B}$ positive definite, then

$$\mathbf{L}_n^{(r)}(\mathbf{Y}_1, \ldots, \mathbf{Y}_n) = \mathbf{B}\mathbf{L}_n^{(r)}(\mathbf{X}_1, \ldots, \mathbf{X}_n) + \mathbf{a},$$

because the $d_{ni}^{(r)}$ are affine-invariant for every $1 \le i \le n$ and for every $r \ge 0$.

5.6.2.1 *Numerical Illustration*

The procedure is illustrated on samples of size $n = 100$ simulated from the multivariate t distribution with 3 degrees of freedom $t_3(\boldsymbol{\theta}, \boldsymbol{\Sigma})$, with

$$\boldsymbol{\theta} = \begin{pmatrix} \theta_1 \\ \theta_2 \\ \theta_3 \end{pmatrix} = \begin{pmatrix} 1 \\ 2 \\ -1 \end{pmatrix} \qquad \boldsymbol{\Sigma} = \begin{bmatrix} 1 & 1/2 & 1/2 \\ 1/2 & 1 & 1/2 \\ 1/2 & 1/2 & 1 \end{bmatrix} \qquad (5.22)$$

and each time, the affine-equivariant trimmed $\mathbf{L}_{n1}$-estimator ($k_n = 15$) and affine-equivariant $\mathbf{L}_{n2}$-estimator were calculated in 10 iterations of the initial estimator. Five thousand replications of the model were simulated and also the mean was computed, for the sake of comparison. Results are summarized in Table 5.2. Figure 5.4 illustrates the distribution of estimated parameters θ_1, θ_2, θ_3 for various iterations of the $\mathbf{L}_{n1}$-estimator and $\mathbf{L}_{n2}$-estimator and compares them with the mean and median.

5.7   Unbiasedness of two-sample nonparametric tests

A frequent practical problem is that we have two data clouds of p-dimensional observations with generally unknown distributions and we wish to test the

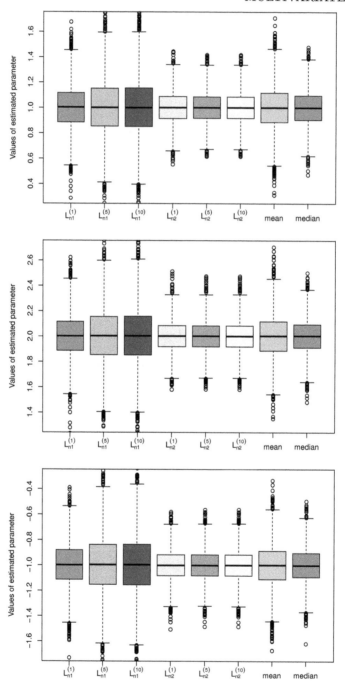

Figure 5.4 *t-distribution: Box-plots of the 5000 estimated values of* $\theta_1(= 1)$ *(top),* $\theta_2(= 2)$ *(middle) and* $\theta_3(= -1)$ *(bottom) for the* $\mathbf{L}_{n1}^{(1)}$, $\mathbf{L}_{n1}^{(5)}$, $\mathbf{L}_{n1}^{(10)}$, $\mathbf{L}_{n2}^{(1)}$, $\mathbf{L}_{n2}^{(5)}$, $\mathbf{L}_{n2}^{(10)}$, *mean and median.*

Table 5.2 *The mean in the sample of 5000 replications of estimators* $\mathbf{L}_{n1}$ *(trimmed) and* $\mathbf{L}_{n2}$, *sample sizes* $n = 100$

i	$\mathbf{L}_{n1}^{(i)}$			$\mathbf{L}_{n2}^{(i)}$		
1	1.003218	2.003116	−0.998004	1.002397	2.002476	−0.999750
2	1.004193	2.004054	−0.996237	1.002318	2.002549	−0.999947
3	1.004316	2.004521	−0.996072	1.002364	2.002706	−1.000094
4	1.004350	2.004901	−0.995854	1.002383	2.002751	−1.000177
5	1.004509	2.005138	−0.995907	1.002397	2.002794	−1.000201
6	1.004307	2.004991	−0.996113	1.002397	2.002803	−1.000209
7	1.004426	2.004994	−0.996239	1.002393	2.002805	−1.000224
8	1.004674	2.005133	−0.996154	1.002380	2.002803	−1.000233
9	1.004805	2.005172	−0.996203	1.002380	2.002806	−1.000234
10	1.004843	2.005148	−0.996136	1.002379	2.002807	−1.000237

hypothesis that they both come from the same distribution, continuous but unknown. This is a nonparametric setting; then we can either utilize the geometry of the data clouds or consider the affine invariance of the testing problem, and look for tests which are invariant or distribution-free. Because the ranks or the signed ranks of geometric entities of data are invariant to many transformations, there are many papers extending the rank methods in the literature. Here belong the tests based on data depths, on the Oja medians, on the multivariate sign functions, and others. Let us refer to the papers of Chaudhuri and Sengupta (1993), Choi and Marden (1997), Marden and Choi (2005), Hallin and Pandaveine (2002,)Hetmansperger et al. (1998), Liu (1988, 1990), Liu and Singh (1993), Oja et al. (1997), Puri and Sen (1971), Randles and Peters (1990), Topchii et al. (2003), Tukey (1975), Zuo and He (2006), among others. An excellent review of the multivariate one-sample, c-sample, independence and ANOVA multivariate rank tests was written by Oja and Randles (2004).

Various permutation tests were proposed by Bickel (1969), Brown (1982), Hall and Tajvidi (2002), Neuhaus and Zhu (1999), Oja (1999), Wellner (1979), among others. Other tests are based on distances between the observations: Let us mention Friedman and Rafsky (1979), Henze (1988), Maa et al. (1996), Rosenbaum (2005), Schilling (1986), and Jurečková and Kalina (2012).

The main components of a testing problem are the hypothesis and the alternative. The hypothesis can be a quite general statement, as that two multivariate populations follow the same distributions. However, our alternative is not simply "the populations follow different probability distributions." Such an alternative does not provide much information, and mathematically, the family of all possible pairs of different distributions is too rich against which we can hardly construct a reasonable test. We should specify in which way the populations differ; it can be a shift in location or in the scale, or in a value of some scalar entity, effect of some covariates, etc. A good test must be unbiased against the considered alternative, i.e., the power of the test of size α against the alternative must not be smaller than α. This problem is often undervalued and people are content with the fact that the test is unbiased against very distant alternatives (is consistent) or that it is unbiased asymptotically for an infinitely increasing number of observations. Because the practice always works with a finite number of observations, we should be interested in the finite-sample behavior of the test. If we are not able to calculate the finite-sample power, we should be at least sure that the test is unbiased against the alternative in mind, at least locally in a neighborhood of the hypothesis. If the rejection probability under the alternative is less than significance level α, we can hardly recommend the test to the experimenter.

5.7.1   Unbiased tests

Let us first consider a general test Φ of the hypothesis

$\{\mathbf{H} :$ distribution F of random vector $\mathbf{X}$ belongs to the set $\mathcal{H}\}$

against the alternative

$\{\mathbf{K} :$ distribution of $\mathbf{X}$ belongs to the set $\mathcal{K}\}$.

If Φ has size α, $0 < \alpha < 1$, where α is the chosen significance level, then $\sup_{F \in \mathcal{H}} \mathrm{E}(\Phi(\mathbf{X})) \leq \alpha$. Test Φ is unbiased, if it satisfies

$$\sup_{F \in \mathcal{H}} \mathrm{E}(\Phi(\mathbf{X})) \leq \alpha \quad \text{and} \quad \inf_{F \in \mathcal{K}} \mathrm{E}(\Phi(\mathbf{X})) \geq \alpha.$$

This is a natural property of a test; the power of a test should not be smaller than the permitted error of the first kind. If there exists a uniformly most powerful test, then it is always unbiased. It is necessary to specify alternatives against which our test is [at least locally] unbiased. That the unbiasedness is not a matter of course follows, e.g., from Jurečková (2002), Jurečková and Milhaud (2003), and Sugiura (1965). Sugiura et al. (2006) even show that while the univariate two-sample Wilcoxon test is always unbiased against one-sided alternatives, it is generally not unbiased against two-sided alternatives. The test is locally unbiased against two-sample alternatives only under some conditions on the hypothetical distribution of observations, e.g., when it is symmetric. Amrhein (1995) demonstrated the same phenomenon for the one-sample Wilcoxon test. Hence, the finite sample unbiasedness of some tests described in the literature is still an open question.

   To illustrate this problem, consider a random vector $\mathbf{X} = (X_1, \ldots, X_n)$ with distribution function $F(\mathbf{x}, \boldsymbol{\theta})$, $\boldsymbol{\theta} \in \boldsymbol{\Theta} \subset \mathbb{R}^p$ and density $f(\mathbf{x}, \boldsymbol{\theta})$ (not necessarily Lebesgue), which has the bounded third derivative in a neighborhood of $\boldsymbol{\theta}_0$ and positively definite Fisher information matrix. We wish to test $\mathbf{H}_0 : \boldsymbol{\theta} = \boldsymbol{\theta}_0$ against alternative $\mathbf{K} : \boldsymbol{\theta} \neq \boldsymbol{\theta}_0$ by test Φ of size α, i.e.,

$E_{\theta_0}[\Phi(\mathbf{X})] = \alpha$. Then we have the following expansion of the power function of Φ (see Jurečková and Milhaud (2003)):

$$E_{\theta_0}\Phi(\mathbf{X}) = \alpha + (\boldsymbol{\theta} - \boldsymbol{\theta}_0)^\top E_{\theta_0}\left\{\Phi(\mathbf{X})\frac{(\dot{\mathbf{f}}(\mathbf{X},\boldsymbol{\theta}_0))}{f(\mathbf{X},\boldsymbol{\theta}_0)}\right\} \tag{5.23}$$

$$+\frac{1}{2}(\boldsymbol{\theta} - \boldsymbol{\theta}_0)^\top E_{\theta_0}\left\{\Phi(\mathbf{X})\frac{[\ddot{\mathbf{f}}(\mathbf{X},\boldsymbol{\theta}_0)]}{f(\mathbf{X},\boldsymbol{\theta}_0)}\right\}(\boldsymbol{\theta} - \boldsymbol{\theta}_0) + \mathcal{O}\left(\|\boldsymbol{\theta} - \boldsymbol{\theta}_0\|^3\right)$$

where

$$(\dot{\mathbf{f}}(\mathbf{x},\boldsymbol{\theta})) = \left(\frac{\partial f(\mathbf{x},\boldsymbol{\theta})}{\partial\theta_1},\ldots,\frac{\partial f(\mathbf{x},\boldsymbol{\theta})}{\partial\theta_p}\right)^\top,$$

$$[\ddot{\mathbf{f}}(\mathbf{x},\boldsymbol{\theta})] = \left[\frac{\partial^2 f(\mathbf{x},\boldsymbol{\theta})}{\partial\theta_j\partial\theta_k}\right]_{j,k=1}^p.$$

The test Φ is locally unbiased if the second term on the right-hand side of (5.23) is non-negative. If θ is a scalar parameter and we consider the one-sided alternative $\mathbf{K}: \theta > \theta_0$, then there always exists an unbiased test. However, if $\boldsymbol{\theta}$ is a vector, then the alternative is only two-sided, and the local unbiasedness of Φ is generally true for a fixed test Φ only under special conditions on f, which we can hardly verify for f unknown. While $E_{\theta_0}\left\{\frac{(\dot{\mathbf{f}}(\mathbf{X},\boldsymbol{\theta}_0))}{f(\mathbf{X},\boldsymbol{\theta}_0)}\right\} = \mathbf{0}$, the following equation

$$E_{\theta_0}\left\{\Phi(\mathbf{X})\frac{(\dot{\mathbf{f}}(\mathbf{X},\boldsymbol{\theta}_0))}{f(\mathbf{X},\boldsymbol{\theta}_0)}\right\} = \mathbf{0} \tag{5.24}$$

is not automatically true. If the test Φ is to be locally unbiased, then it should satisfy (5.24), because otherwise the second term in (5.23) can be negative for some $\boldsymbol{\theta}$ and the power can be less than α. This was observed already by Grose and King (1991), who imposed condition (5.24) when they constructed a locally unbiased two-sided version of the Durbin-Watson test.

If we represent the multivariate observations by suitable scalar values on which we construct the test, then the test is always locally unbiased against the alternatives, one-sided in these scalars. Such test is useful if the monotonicity of the scalar entities is meaningful with respect to the original observations. From this point of view, the most useful scalar characteristics of observations are distances of various types between the observational points. Besides that, if the distribution of the scalar characteristic depends on some real parameter, then as one-sided alternatives we can consider the families of distributions of the scalars, stochastically monotone in this parameter. Some tests are illustrated in the following text.

5.7.2   Some multivariate two-sample tests

Consider two independent samples $\mathcal{X} = (\mathbf{X}_1, \ldots, \mathbf{X}_m)$ and $\mathcal{Y} = (\mathbf{Y}_1, \ldots, \mathbf{Y}_n)$ from two p-variate populations with continuous distribution functions $F^{(p)}$ and $G^{(p)}$, with the respective means and dispersion matrices $\boldsymbol{\mu}_1, \boldsymbol{\mu}_2, \boldsymbol{\Sigma}_1, \boldsymbol{\Sigma}_2$. The problem is to test the hypothesis $\mathbf{H}_0 : F^{(p)} \equiv G^{(p)}$ (along with $\boldsymbol{\mu}_1 = \boldsymbol{\mu}_2, \boldsymbol{\Sigma}_1 = \boldsymbol{\Sigma}_2$) against an alternative $\mathbf{H}_1$ where either $(\boldsymbol{\mu}_1, \boldsymbol{\Sigma}_1) \neq (\boldsymbol{\mu}_2, \boldsymbol{\Sigma}_2)$ or where $F^{(p)}$ and $G^{(p)}$ are not of the same functional form. Denote $(\mathbf{Z}_1, \ldots, \mathbf{Z}_N)$ as the pooled sample with $\mathbf{Z}_i = \mathbf{X}_i$, $i = 1, \ldots, m$ and $\mathbf{Z}_{m+j} = \mathbf{Y}_j$, $j = 1, \ldots, n$, $m + n = N$. The hypothesis and the alternative are invariant under affine transformations

$$\mathcal{G} : \{\mathbf{Z} \to \mathbf{a} + \mathbf{B}\mathbf{Z}\} \text{ with } \mathbf{a} \in \mathbb{R}^p \text{ and nonsingular } p \times p \text{ matrix } \mathbf{B}. \quad (5.25)$$

The invariant tests depend on the data only by means of a *maximal invariant* of $\mathcal{G}$ (Lehmann (1997). Obenchain (1971) showed that the maximal invariant with respect to $\mathcal{G}$ is

$$\mathbf{T}(\mathbf{Z}_1, \ldots, \mathbf{Z}_N) = \left[(\mathbf{Z}_i - \bar{\mathbf{Z}}_N)^\top \mathbf{V}_N^{-1}(\mathbf{Z}_j - \bar{\mathbf{Z}}_N)\right]_{i,j=1}^N, \quad (5.26)$$

$$\text{where} \quad \bar{\mathbf{Z}}_N = \frac{1}{N} \sum_{i=1}^N \mathbf{Z}_i, \ \mathbf{V}_N = \sum_{i=1}^N (\mathbf{Z}_i - \bar{\mathbf{Z}}_N)(\mathbf{Z}_i - \bar{\mathbf{Z}}_N)^\top.$$

Then $\mathbf{T}(\mathbf{Z}_1, \ldots, \mathbf{Z}_N)$ is the projection matrix associated with the space spanned by the columns of the matrix $\left[\mathbf{Z}_1 - \bar{\mathbf{Z}}_N, \ldots, \mathbf{Z}_N - \bar{\mathbf{Z}}_N\right]$.

Example 5.1 *The well-known two-sample **Hotelling T^2** test is based on the criterion*

$$\mathcal{T}_{mn}^2 = (\bar{\mathbf{X}}_m - \bar{\mathbf{Y}}_n)^\top \mathbf{V}_N^{-1}(\bar{\mathbf{X}}_m - \bar{\mathbf{Y}}_n). \quad (5.27)$$

The test is invariant with respect to $\mathcal{G}$ and is optimal unbiased against two-sample normal alternatives with $\boldsymbol{\mu}_1 \neq \boldsymbol{\mu}_2$ and $\boldsymbol{\Sigma}_1 = \boldsymbol{\Sigma}_2$. Its asymptotic distribution under $\mathbf{H}_0$, when both sample sizes m, n tend to infinity, does not depend on the normality. If $m, n \to \infty$ and $\frac{m}{n} \to 1$, then the asymptotic distribution of $\mathcal{T}_{mn}^2$ does not change even when $\boldsymbol{\Sigma}_1 \neq \boldsymbol{\Sigma}_2$, but only in this case (see Lehmann (1997)). The finite sample unbiasedness of the Hotelling test is not guaranteed if the distributions are not normal.

Example 5.2 *Liu and Singh rank sum test.* *A test of Wilcoxon type, proposed by Liu and Singh (1993), is based on the ranks of depths of the data. Its asymptotic distributions under the hypothesis and under general alternative distributions F, G of depths was derived by Zuo and He (2006).*

Let $D(\mathbf{y}; H)$ denote a depth function of a distribution H evaluated at point $\mathbf{y} \in \mathbb{R}^p$. Liu and Singh (1993) considered a parameter, called a quality index, defined as

$$Q(F^{(p)}, G^{(p)}) = \int R(\mathbf{y}; F^{(p)}) dG^{(p)}(\mathbf{y})$$

$$= \mathbb{P}\left\{ D(\mathbf{X}; F^{(p)}) \leq D(\mathbf{Y}; F^{(p)}) \Big| \mathbf{X} \sim F^{(p)}, \mathbf{Y} \sim G^{(p)} \right\}$$

where $R(\mathbf{y}; F^{(p)}) = \mathbb{P}_F\left(D(\mathbf{X}; F^{(p)}) \le D(\mathbf{y}; F^{(p)})\right),\ \mathbf{y} \in \mathbb{R}^p$, *and showed that, if* $D(\mathbf{X}; F^{(p)})$ *has a continuous distribution, then* $Q(F^{(p)}, F^{(p)}) = \frac{1}{2}$. *Then they tested the hypothesis* $Q(F^{(p)}, G^{(p)}) = \frac{1}{2}$ *against the alternative* $Q(F^{(p)}, G^{(p)}) \ne \frac{1}{2}$, *using the Wilcoxon-type criterion based on the empirical distribution functions* F_m, G_n *of samples of sizes* m, n :

$$Q(F_m, G_n) = \int R(\mathbf{y}; F_m) dG_n(\mathbf{y}) = \frac{1}{n} \sum_{j=1}^{n} R(\mathbf{Y}_j; F_m).$$

If the distribution of depths is symmetric under $F^{(p)} \equiv G^{(p)}$, *then the test rejecting provided* $|Q(F_m, G_n) - \frac{1}{2}| \ge C_{\alpha/2}$ *is locally unbiased against* $Q(F^{(p)}, G^{(p)}) \ne \frac{1}{2}$. *Under general distribution of depths, only the one-sided test with the critical region*

$$Q(F_m, G_n) - \frac{1}{2} > C_\alpha$$

is unbiased against the one-sided alternative $Q(F^{(p)}, G^{(p)}) > \frac{1}{2}$; *however, this alternative, one-sided in depths, has a difficult interpretation with respect to distributions* $F^{(p)}, G^{(p)}$ *of original observations* $\mathbf{X}$ *and* $\mathbf{Y}$. *Generally the test is not finite-sample unbiased against* $F \ne G$, *not even locally. The unbiasedness can be guaranteed only in some cases, e.g., if the hypothetical distribution of depths is symmetric.*

If the distribution of depths is symmetric under $F^{(p)} \equiv G^{(p)}$, *then the test rejecting for* $|Q(F_m, G_n) - \frac{1}{2}| \ge C_{\alpha/2}$ *is locally unbiased against* $Q(F^{(p)}, G^{(p)}) \ne \frac{1}{2}$. *Generally, if the distribution of depths is not symmetric, only the one-sided test with the critical region*

$$Q(F_m, G_n) - \frac{1}{2} > C_\alpha$$

is unbiased against the one-sided alternative $Q(F^{(p)}, G^{(p)}) > \frac{1}{2}$. *However, it is difficult to explain the alternative, one-sided in depths, in distributions* $F^{(p)}, G^{(p)}$ *of original observations* $\mathbf{X}$ *and* $\mathbf{Y}$.

Example 5.3 *Rank test based on distances of observations.* Jurečková *and* Kalina (2012) *considered the tests based on ranks of distances between observations. The alternatives claiming that some distances are greater than others are one-sided, hence the tests are unbiased.*

Denote

$$\mathcal{Z} = (\mathbf{Z}_1, \ldots, \mathbf{Z}_N) = (\mathbf{X}_1, \ldots, \mathbf{X}_m, \mathbf{Y}_1, \ldots, \mathbf{Y}_n)$$

as the pooled sample, $N = m + n$, $\mathbf{Z}_i \in \mathbb{R}^p$. *Choose a distance* $L = L(\cdot, \cdot)$ *in* $\mathbb{R}^p$, *taking on nonnegative real values, and consider the matrix of distances* $\mathbb{L}_N = [\ell_{ik}]_{i,k=1}^N$, *where* $\ell_{ik} = L(\mathbf{Z}_i, \mathbf{Z}_k)$. *The rank tests based on* $\mathbb{L}_N$ *can be constructed in various ways, with various types of invariance.*

For every fixed i and under fixed $\mathbf{X}_i$, $1 \le i \le m$, we can consider the distances $\{\ell_{ik} = L(\mathbf{X}_i, \mathbf{Z}_k),\ k = 1,\ldots,N,\ k \ne i\}$. Then, conditionally given $\mathbf{X}_i$, the vector $\{\ell_{ik},\ k = 1,\ldots,m,\ k \ne i\}$ is a random sample from a population with a distribution function $F(z|\mathbf{X}_i) = F$ (say), while $\{\ell_{ik},\ k = m+1,\ldots,N\}$ is a random sample from a population with a distribution function $G(z|\mathbf{X}_i) = G$. Assuming that the distribution functions F and G are absolutely continuous, we work with the ranks

$$\mathcal{R}_i = (R_{i1},\ldots,R_{i,i-1},R_{i,i+1},\ldots,R_{iN})$$

the ranks of $\ell_{ik},\ k = 1,\ldots,N,\ k \ne i$. Every two-sample rank test will depend only on the ordered ranks $R_i^{(m+1)} < \ldots < R_i^{(N)}$ of the second sample. Especially, if $L(\mathbf{X}_i, \mathbf{Z}_k) = \|\mathbf{X}_i - \mathbf{Z}_k\|,\ k = 1,\ldots,N,\ k \ne i$, where $\|\cdot\|$ is the Euclidean distance, then the test based on their ranks will be invariant to the shift in location, but not affine invariant. The linear (conditional) rank test is based on the linear rank statistic

$$S_{iN} = N^{-1/2} \sum_{k=m+1}^{N} a_N(R_{ik}) \tag{5.28}$$

with the scores $a_N(1),\ldots,a_N(N-1)$ generated by a nondecreasing score function φ.

The criteria S_{iN} are equally distributed for $i = 1,\ldots,m$, under the hypothesis and under the alternatives, though not independent. Using only a single S_{iN} would be a loss of information, so we look for a convenient combination of $S_{1N},\ldots,S_{mN}$. The test based on a single S_{iN} is a standard rank test, e.g., Wilcoxon, conditionally given $\mathbf{X}_i$, and thus easy to perform. A randomization of $S_{1N},\ldots,S_{mN}$, keeps the simple structure of the test and is thus easy to perform; it leads to the following criterion $S^{(N)}$:

$$\mathbf{P}(S^{(N)} = S_{iN}) = \frac{1}{m}, \quad i = 1,\ldots,m \tag{5.29}$$

where the randomization in (5.29) is independent of the set of observations $\mathcal{Z}$. The following identity is true for any C :

$$\mathbf{P}\left(S^{(N)} > C\right) = \frac{1}{m} \sum_{i=1}^{m} \mathbf{P}\left(S_{iN} > C\right),$$

and the test rejects $\mathbf{H}_0$ for $\alpha \in (0,1)$ if $S^{(N)} > C_\alpha$, eventually it rejects with probability $\gamma \in (0,1)$ if $S^{(N)} = C_\alpha$, where

$$\mathbf{P}_{\mathbf{H}_0}\left(S^{(N)} > C_\alpha\right) + \gamma \mathbf{P}_{\mathbf{H}_0}\left(S^{(N)} = C_\alpha\right) = \alpha.$$

Example 5.4 *In Example 5.3 the distances from one point at a time are used. Let us now consider a two-sample test statistic based on ranks that uses the whole matrix of distances $\mathbb{L}_N = [\ell_{ik}]_{i,k=1}^N$. Consider a matrix $\mathbf{R}_N = [r_{ik}]_{i,k=1}^N$ with $i-$th row*

$$\mathbf{r}_i = (R_{i1}, \ldots, R_{i,i-1}, 0, R_{i,i+1}, \ldots, R_{iN}), \quad i = 1, \ldots, N$$

where

$$(R_{i1}, \ldots, R_{i,i-1}, R_{i,i+1}, \ldots, R_{iN})$$

are the ranks of ℓ_{ik}, $k = 1, \ldots, N$, $k \neq i$. Note that the matrix $\mathbf{R}_N$ is not symmetric. The test is based on the statistic

$$\mathcal{N} = \left[\frac{1}{mn} \sum_{i=1}^m \sum_{j=m+1}^N (r_{ij} + r_{ji}) - \frac{1}{m^2} \sum_{i=1}^m \sum_{j=1}^m r_{ij} - \frac{1}{n^2} \sum_{i=m+1}^N \sum_{j=m+1}^N r_{ij} \right]^{\frac{1}{2}}$$

$$(5.30)$$

which can be, in case of $m = n$, rewritten as

$$\mathcal{N} = \left[\frac{2}{n^2} \sum_{i=1}^n \sum_{j=n+1}^{2n} (r_{ij} + r_{ji}) - 2(2n-1) \right]^{\frac{1}{2}}.$$

A permutation strategy is then used to estimate the p-value of the test. The elements of the vector $(\mathbf{Z}_1, \ldots, \mathbf{Z}_N)$ are permuted several times and the p-value of the test is calculated as the proportion of permutations where the statistic $\mathcal{N}$ is greater than or equal to $\mathcal{N}$ for the observed data.

The formula (5.30) is inspired by a dissimilarity measure of two clusters when applying Ward's method of hierarchical cluster analysis. When the measure is expressed in terms of interpoint distances it is equivalent to (5.30).

The same statistic, but not based on ranks, is e.g., used in Klebanov (2005) as a two-sample test criterion based on $\mathcal{N}-$distances.

The empirical powers of the Hotelling T^2 (Example 5.1), the Wilcoxon type (Example 5.3) and the $\mathcal{N}-$distance type (Example 5.4) two-sample tests are compared under bivariate normal and Cauchy distributions with various parameters. Both Wilcoxon type and $\mathcal{N}-$distance type tests are based on the ranks of the Euclidean interpoint distances. The simulations are based on 1000 replications. For the $\mathcal{N}-$distance type permutation test, 1000 permutations are used in each replication. The Hotelling test distinguishes well two normal samples differing fairly in locations, even if they also differ in scales. In some situations, Wilcoxon even competes well with Hotelling. However, the $\mathcal{N}-$distance type test has the best overall performance. This is illustrated in Table 5.3. Table 5.4 presents the empirical powers of the tests comparing two samples from the bivariate Cauchy distributions. In this case, the Wilcoxon test is far more powerful than the Hotelling test, already under a small shift.

The Hotelling test fails completely if $\boldsymbol{\mu} = 0$ but $\sigma \neq 1$, while the Wilcoxon still distinguishes the samples well. Wilcoxon dominates Hotelling also in other situations. However, the $\mathcal{N}$−distance type test is even more powerful than the Wilcoxon test in nearly all cases.

Table 5.3 *Powers of two-sample Hotelling T^2 test (H), two-sample Wilcoxon test (W) and two-sample $\mathcal{N}$−distance type test (N) based on distances for various $m = n$ and $\alpha = 0.05$. The first sample always has $\mathcal{N}_2(\boldsymbol{\mu}_1, \boldsymbol{\Sigma}_1)$ distribution with $\boldsymbol{\mu}_1 = (0,0)^\top$ and $\boldsymbol{\Sigma}_1 = \mathrm{Diag}\{1,1\}$. The second sample has $\mathcal{N}_2(\boldsymbol{\mu}_2, \boldsymbol{\Sigma}_2)$ with various $\boldsymbol{\mu}_2, \boldsymbol{\Sigma}_2$ specified in the first column.*

Second sample	Test	$m = n = 10$	$m = n = 100$	$m = n = 1000$
	H	0.054	0.042	0.052
$\boldsymbol{\mu}_2 = (0,0)^T$	W	0.044	0.045	0.041
$\boldsymbol{\Sigma}_2 = \mathrm{Diag}\{1,1\}$	N	0.046	0.029	0.054
	H	0.098	0.401	1.000
$\boldsymbol{\mu}_2 = (0.2, 0.2)^T$	W	0.064	0.160	0.673
$\boldsymbol{\Sigma}_2 = \mathrm{Diag}\{1,1\}$	N	0.087	0.359	1.000
	H	0.062	0.049	0.042
$\boldsymbol{\mu}_2 = (0,0)^T$	W	0.106	0.633	0.983
$\boldsymbol{\Sigma}_2 = \mathrm{Diag}\{2,2\}$	N	0.089	0.749	1.000
	H	0.042	0.108	0.712
$\boldsymbol{\mu}_2 = (0.1, 0.1)^T$	W	0.064	0.354	0.831
$\boldsymbol{\Sigma}_2 = \mathrm{Diag}\{1.5, 1.5\}$	N	0.065	0.309	1.000

Remark 5.1 *The Mahalanobis distances*

$$(\mathbf{X}_i - \mathbf{Z}_k)^\top \mathbf{V}_N^{-1}(\mathbf{X}_i - \mathbf{Z}_k) \quad or \quad (\mathbf{X}_i - \mathbf{Z}_k)^\top (\mathbf{V}_N^0)^{-1}(\mathbf{X}_i - \mathbf{Z}_k), \ k \neq i$$

$$\mathbf{V}_N = \sum_{i=1}^N (\mathbf{Z}_i - \bar{\mathbf{Z}}_N)(\mathbf{Z}_i - \bar{\mathbf{Z}}_N)^\top, \ \mathbf{V}_N^0 = \sum_{i=1}^N \mathbf{Z}_i \mathbf{Z}_i^\top$$

are not independent, but under $\mathbf{H}_0$ they have exchangeable distributions. Hence, under $\mathbf{H}_0$ the distribution of their ranks is distribution-free, and they are also affine invariant. However, the invariant tests based on ranks of Mahalanobis distances are still an open problem; their structure would be more complex than if the tests are based on simple distances.

5.8 Problems and complements

5.1 $\mathbf{X}$ is spherically symmetric about $\boldsymbol{\theta}$ if and only if $\|\mathbf{X} - \boldsymbol{\theta}\|$ and $\frac{\mathbf{X}-\boldsymbol{\theta}}{\|\mathbf{X}-\boldsymbol{\theta}\|}$ are independent (Dempster (1969)).

5.2 Let T_1 and T_2 be two competing estimators of $\boldsymbol{\theta}$ and $L(\cdot, \cdot)$ be a loss function. Denote

$$\mathcal{P}(\mathbf{T}_1, \mathbf{T}_2; \boldsymbol{\theta}) = P_\theta\{L(\mathbf{T}_1, \boldsymbol{\theta}) < L(\mathbf{T}_2, \boldsymbol{\theta})\} + \tfrac{1}{2}P_\theta\{L(\mathbf{T}_1, \boldsymbol{\theta}) = L(\mathbf{T}_2, \boldsymbol{\theta})\}.$$

Table 5.4 *Powers of two-sample Hotelling T^2 test (H), two-sample Wilcoxon test (W) and two-sample $\mathcal{N}$–distance type test (N) based on distances for various $m = n$ and $\alpha = 0.05$. The first sample X always has the two-dimensional Cauchy distribution. The second sample Y is obtained as $Y = \mu + \sigma Y^*$, where Y^* is generated as a two-dimensional Cauchy distribution independent on X. Values of m, n, μ and σ are specified in the first column.*

Second sample	Test	$m = n = 10$	$m = n = 100$	$m = n = 1000$
	H	0.022	0.015	0.014
$\mu = (0,0)^T$	W	0.044	0.050	0.061
$\sigma = 1$	N	0.058	0.047	0.066
	H	0.016	0.024	0.016
$\mu = (0.2, 0.2)^T$	W	0.043	0.096	0.315
$\sigma = 1$	N	0.056	0.127	0.852
	H	0.018	0.021	0.014
$\mu = (0,0)^T$	W	0.158	0.680	0.961
$\sigma = 2$	N	0.119	0.889	1.000
	H	0.058	0.054	0.039
$\mu = (1,1)^T$	W	0.150	0.702	0.885
$\sigma = 2$	N	0.234	0.996	1.000

Then T_1 is said to be Pitman-closer to θ than T_2 if $\mathcal{P}(T_1, T_2; \theta) \geq \frac{1}{2}$, with a strict inequality for at least some θ. The estimator T_1 is called the Pitman closest one of θ in the family $\mathcal{C}$, if $\mathcal{P}(T_1, T_2; \theta) \geq \frac{1}{2} \ \forall \theta$ and all $T \in \mathcal{C}$. (See Sen (1994).)

5.3 Let $x_1, \ldots, x_n$ be distinct points in $\mathbb{R}^p$. Consider the minimization of

$$C(y) = \sum_{i=1}^{n} \eta_i \| y - x_i \| \tag{5.31}$$

with respect to $y \in \mathbb{R}^p$, where $\eta_1, \ldots, \eta_n$ are chosen positive weights. The solution of (5.31) is called the L_1-*multivariate median*. If $C(y)$ is strictly convex in $\mathbb{R}^p$ and $x_1, \ldots, x_n$ are not collinear, then the minimum is uniquely determined. If $x_1, \ldots, x_n$ lie in a straight line, the minimum is achieved at any one-dimensional median and may not be unique (Vardi and Zhang (2000)).

5.4 Let $x_1, \ldots, x_n$ and $\eta_1, \ldots, \eta_n$ be the same as in Problem **5.3** and $\xi_i = \eta_i / (\sum_{j=1}^{n} \eta_j)$, $i = 1, \ldots, n$. Denote

$$D(y) = \begin{cases} 1 - \|\bar{e}(y)\| & \text{if } y \text{ differs from } \{x_1, \ldots, x_n\}, \\ 1 - (\|\bar{e}(y)\| - \xi_k) & \text{if } y = x_k, \text{ for some } k \end{cases}$$

where $e_i(y) = \frac{y - x_i}{\|y - x_i\|}$, $i = 1, \ldots, n$ and $\bar{e}(y) = \sum_{\{i : x_i \neq y\}} \xi_i e_i(y)$. The function $\bar{e}(y)$ is the *spatial rank function* considered by Möttönen and Oja

(1995) and by Marden (1998). Chaudhuri (1996) considered the corresponding multivariate quantiles.

Verify that $\|\bar{\mathbf{e}}(\mathbf{y})\| \leq \sum_{\{k:\mathbf{x}_k \neq \mathbf{y}\}} \xi_k \leq 1$, hence $0 \leq D(\mathbf{y}) \leq 1$ and $\lim_{\|\mathbf{y}\| \to \infty} D(\mathbf{y}) = 0$. Moreover, $D(\mathbf{x}_k) = 1$ if $\xi_k \geq \frac{1}{2}$, i.e., the point $\mathbf{x}_k$ is the L_1-multivariate median if it possesses half of the total weight (Vardi and Zhang (2000)).

Chapter 6

Large sample and finite sample behavior
of robust estimators

6.1 Introduction

Robust estimators are non-linear functions of observations, often implicitly defined. It is often difficult to derive their distribution functions under a finite number of observations. If we are not able to derive the finite-sample distribution in a compact form, we take recourse to a limit distribution function of an estimator and/or criterion under the condition that the number of observations n is infinitely increasing (*asymptotic distribution*).

The asymptotic distribution is often normal, and its variance is considered as an important characteristic of the estimator. Because the robust estimators are not linear combinations of independent random variables, their asymptotic distributions cannot be derived by a straightforward application of the central limit theorem. In the first step we approximate the sequence $\sqrt{n}(T_n - T(F))$ by a linear combination of independent summands, and to it we then apply the central limit theorem. In the literature, this approximation is known as an *asymptotic representation* of T_n. If the functional $T(P)$ is Fréchet differentiable, then the asymptotic representation follows from the expansion (1.18); it can be written in the form

$$\sqrt{n}(T_n - T(F)) = \frac{1}{n}\sum_{i=1}^{n} IF(X_i, T, F) + R_n \qquad (6.1)$$

where $IF(x, T, P)$ is the influence function of $T(F)$ and the remainder term is asymptotically negligible as $n \to \infty$, i.e., $R_n = o_p(1)$. An asymptotic representation of estimator T_n can be derived even if it is not Fréchet differentiable, when the score functions (ψ, J, φ), of M-, L- and R-estimators and distribution functions F of the model errors are sufficiently smooth or have other convenient properties. The asymptotic theory of robust and nonparametric statistical procedures leads to interesting and challenging mathematical problems, and a host of excellent papers and books were devoted to their solutions. However, we should be cautious and keep in mind that the asymptotic solution is rather a tool for how to get over our ignorance of the precise finite-sample properties. We shall see that the asymptotics can sometimes distort the true behavior of the procedure.

The asymptotic properties of robust estimators and their asymptotic relations were studied in detail in several excellent monographs; we refer to Bickel et al. (1993), Dodge and Jurečková (2000), Field and Ronchetti (1990), Hampel et al. (1986, 2005), Hettmansperger and McKean (2011), Huber (1981), Huber and Ronchetti (2009), Jurečková and Sen (1996), Jurečková, Sen and Picek (2013), Koenker (2005), Koul (2002), Lehmann (1999), Maronna et al. (2006), Oja (2010), Rieder (1994), Serfling (1980), Shevlyakov and Vilchevski (2001), Shorack and Wellner (1986), Shorack (2000), Sen (1981), van der Vaart and Wellner (1996), Witting and Müller-Funk (1995), among others. Regarding that, in this chapter we shall give only a brief outline of basic asymptotic results for M-, L- and R-estimators, mostly of the location parameter. We shall devote more space to the asymptotic relations and equivalence of various estimators, which can be of independent interest.

If an estimator T_n admits a representation (6.1), then it is asymptotically normally distributed as $n \to \infty$ in the sense that

$$\mathcal{L}\left\{\sqrt{n}(T_n - T(F))\right\} \to \mathcal{N}(0, \sigma_F^2) \tag{6.2}$$

where $\sigma_F^2 = \mathrm{E}_F(IF(X, T, F))^2$. We shall describe the asymptotic representations and asymptotic distributions of M-, L- and R-estimators of a scalar parameter, whose influence functions we have derived earlier. The representations are parallel in the linear regression model. Comparing the asymptotic representations enables us to discover the mutual relations of various robust estimators, namely to see when the pairs of M- and R-estimators are asymptotically equivalent; similarly for other pairs (Sections 6.2–6.5). Of most interest is the behavior of robust estimators under contaminated distribution of model errors, more precisely, when the distribution is not the assumed F, but $(1 - \varepsilon)F + \varepsilon H$ with an unknown distribution function H. This can be interpreted so that $100(1 - \varepsilon)\%$ observations come from the "correct" distribution, while $100\varepsilon\%$ observations are contaminated and have an unknown distribution. With unknown H, we can only look for a minimax M-estimator (or R- and L-estimator) with minimal asymptotic variance for a least favorable distribution of the contaminated model (Section 6.6). The asymptotic results will be formulated mostly without proofs, with a reference to the literature.

We shall also pay due attention to the behavior of estimators under finite number n of observations. There can be considerable differences in the behavior under finite or infinite samples. The estimators are often asymptotically normally distributed, hence asymptotically admissible, while their distribution can be heavy-tailed under finite n and they can be inadmissible, even they cannot be Bayesian (Section 6.7).

If an estimator is defined implicitly as a solution of a system of equations, then its calculation can be computationally difficult. Then it can be approximated by one or two Newton-Raphson iterations, started with a suitable initial estimator. Already the first iteration can be asymptotically equivalent to the solution of equations, but we cannot neglect the choice of the initial estimator. The approximation typically inherits the finite-sample breakdown point and tail-behavior from the initial estimator, while it retains only the asymptotic properties. It is even possible to combine a poor initial estimator with good iterations and combine a good breakdown point with good asymptotic efficiency (Section 6.8). Finally, we can use a convex combinaison of two estimating procedures if we cannot prefer either of them; this combination can even be adaptive, driven by the observations (Section 6.9).

6.2   M-estimators

6.2.1   M-estimator of general scalar parameter

Let $\{X_i,\ i = 1, 2, \ldots, \}$ be a sequence of independent observations with a joint distribution function $F(x, \theta)$, $\theta \in \Theta$, where Θ is an open interval of $\mathbb{R}^1$. The M-estimator of θ is a solution of the minimization

$$\sum_{i=1}^{n} \rho(X_i, \theta) = \min, \ \theta \in \Theta$$

Assume that $\rho(x, \theta)$ is absolutely continuous in θ with derivative $\psi(x, \theta) = \frac{\partial}{\partial \theta} \rho(x, \theta)$. If $\psi(x, \theta)$ is continuous in θ, then we look for the M-estimator T_n among the roots of the equation

$$\sum_{i=1}^{n} \psi(X_i, \theta) = 0 \qquad (6.3)$$

If the function $\mathrm{E}\{\rho(X, t)|\theta\}$ has a unique minimum at $t = \theta$ (hence, the functional is Fisher consistent) and some other conditions on either the smoothness of $\psi(x, \theta)$ or the smoothness of $F(x, \theta)$ are satisfied, then there exists a sequence $\{T_n\}$ of roots of the Equation (6.3) such that, as $n \to \infty$,

$$\sqrt{n}(T_n - \theta) = \mathcal{O}_p(1)$$

$$\sqrt{n}(T_n - \theta) = \frac{1}{\sqrt{n}\gamma(\theta)} \sum_{i=1}^{n} \psi(X_i, \theta) + \mathcal{O}_p(n^{-1/2}) \qquad (6.4)$$

$$\text{where } \gamma(\theta) = \mathrm{E}_\theta(\dot{\psi}(X, \theta)), \ \dot{\psi}(x, \theta) = \frac{\partial}{\partial \theta} \psi(x, \theta)$$

This further implies that $\sqrt{n}(T_n - \theta)$ has asymptotically normal distribution $\mathcal{N}(0, \sigma^2(\psi, F))$ where

$$\sigma^2(\psi, F)) = \frac{\mathrm{E}_\theta(\psi^2(X, \theta))}{\gamma^2(\theta)} \qquad (6.5)$$

6.2.2   M-estimators of location parameter

Let $X_1, X_2, \ldots$ be independent observations with joint distribution function $F(x - \theta)$. The M-estimator of θ is a solution of the minimization

$$\sum_{i=1}^{n} \rho(X_i - \theta) = \min, \quad \theta \in \mathbb{R}^1$$

Assume that $\rho(x)$ is absolutely continuous with derivative $\psi(x)$ and that the function $h(t) = \int_{\mathbb{R}} \rho(x - t) dF(x)$ has a unique minimum at $t = 0$. If ψ is absolutely continuous with derivative ψ' and $\gamma = \int \psi'(x) dF(x) > 0$, then there exists a sequence $\{T_n\}$ of roots of the equation $\sum_{i=1}^{n} \psi(X_i - t) = 0$ such that

$$\sqrt{n}(T_n - \theta) = \mathcal{O}_p(1)$$

$$\sqrt{n}(T_n - \theta) = \frac{1}{\sqrt{n}\gamma} \sum_{i=1}^{n} \psi(X_i - \theta) + \mathcal{O}_p(n^{-1/2}) \tag{6.6}$$

$$P_\theta\left(\sqrt{n}(T_n - \theta) \le x\right) \to \Phi\left(\frac{x}{\sigma(\psi, F)}\right) \quad \text{as } n \to \infty$$

where $\sigma^2(\psi, F) = \gamma^{-2} \int_{\mathbb{R}} \psi^2(x) dF(x)$ and Φ is the standard normal distribution function $\mathcal{N}(0, 1)$. If F has an absolutely continuous density f with derivative f' and finite Fisher information $\mathcal{I}(F) = \int [f'(x)/f(x)]^2 dF(x) > 0$, then, under a special choice $\rho(x) = -\log f(x)$, the M-estimator coincides with the maximal likelihood estimator of θ, and its asymptotic variance attains the Rao-Cramér lower bound $1/\mathcal{I}(F)$.

    If $\psi(x)$ has points of discontinuity, then we should assume that the distribution function F has two derivatives f, f' in their neighborhoods. The M-estimator is uniquely determined by relations (3.12), provided ψ is nondecreasing. The solution T_n of the minimization $\sum_{i=1}^{n} \rho(X_i - \theta) := \min$ is not necessarily a root of the equation $\sum_{i=1}^{n} \psi(X_i - \theta) = 0$, but it always satisfies

$$n^{-1/2} \sum_{i=1}^{n} \psi(X_i - T_n) = \mathcal{O}_p(n^{-1/2}) \quad \text{as } n \to \infty \tag{6.7}$$

and

$$\sqrt{n}(T_n - \theta) = \frac{1}{\sqrt{n}\gamma'} \sum_{i=1}^{n} \psi(X_i - \theta) + \mathcal{O}_p(n^{-1/4})$$

$$\gamma' = \int_{\mathbb{R}} f(x) d\psi(x), \tag{6.8}$$

$$P_\theta\left(\sqrt{n}(T_n - \theta) \le x\right) \to \Phi\left(\frac{x}{\sigma(\psi, F)}\right)$$

where $\sigma^2(\psi, F) = (\gamma')^{-2} \int_{\mathbb{R}} \psi^2(x) dF(x)$ and Φ is a distribution function of the normal distribution $\mathcal{N}(0, 1)$.

Finally, let T_n^* be a scale equivariant M-estimator of θ, studentized by the scale statistic S_n such that $S_n \to S(F) = S$ as $n \to \infty$. Then, if both ψ and F are absolutely continuous and smooth, T_n^* admits the asymptotic representation

$$T_n^* - \theta = \frac{1}{n\gamma_1} \sum_{i=1}^{n} \psi\left(\frac{X_i - \theta}{S}\right) - \frac{\gamma_2}{\gamma_1}\left(\frac{S_n}{S} - 1\right) + \mathcal{O}_p(n^{-1})$$

where $\gamma_1 = \frac{1}{S} \int \psi'\left(\frac{x}{S}\right) dF(x)$ and $\gamma_2 = \frac{1}{S} \int x\psi'\left(\frac{x}{S}\right) dF(x)$. However, we cannot separate T_n^* from S_n, unless we impose additional conditions on ψ, F and on S_n. Hence $\sqrt{n}(T_n^* - \theta)$ is not asymptotically normally distributed itself, but we have

$$\mathcal{L}\left\{\sqrt{n}\left[\gamma_1(T_n^* - \theta) + \gamma_2\frac{S_n}{S}\right]\right\} \to \mathcal{N}(0, \sigma^2)$$

where $\sigma^2 = \int \psi^2\left(\frac{x}{S}\right) dF(x)$. In the symmetric case, where $\psi(-x) = \psi(x)$ and $F(x) + F(-x) = 1$, $x \in \mathbb{R}$, is $\gamma_2 = 0$ and $\sqrt{n}(T_n^* - \theta)$ is asymptotically normal.

6.3   *L*-estimators

Let $X_1, X_2, \ldots$, be independent observations with a joint distribution function F. The L-estimator of type I is a linear combination of order statistics $T_n = \sum_{i=1}^{n} c_{ni} X_{n:i}$ with the coefficients generated by the weight function J according to (3.63) or (3.64). Among them, we shall concentrate on robust members, i.e., on the trimmed L-estimators with $J(u) = 0$ for $0 \le u < \alpha$ and $1 - \alpha < u \le 1$, $0 < \alpha < \frac{1}{2}$. To derive the asymptotic representation of such L-estimators, we should assume that the distribution function F is continuous almost everywhere and that F and J have no joint point of discontinuity. More precisely, if J is discontinuous at u, then $F^{-1}(u)$ should be Lipschitz in a neighborhood of u. Under these conditions, as $n \to \infty$,

$$\sqrt{n}(T_n - T(F)) = n^{-1/2} \sum_{i=1}^{n} \psi_1(X_i) + \mathcal{O}_p(n^{-1/2})$$

where
$$T(F) = \int_0^1 J(u) F^{-1}(u) du \tag{6.9}$$

and
$$\psi_1(x) = -\int_{\mathbb{R}} \{I[y \ge x] - F(y)\} J(F(y)) dy, \ x \in \mathbb{R}$$

Moreover, $\sqrt{n}(T_n - T(F))$ has asymptotically normal distribution $\mathcal{N}\left(0, \sigma^2(J, F)\right)$, where

$$\sigma^2(J, F) = \int_{\mathbb{R}} \psi_1^2(x) dF(x)$$

$$= \int_{-\infty}^{\infty} \int_{-\infty}^{\infty} J(F(x)) J(F(y)) [F(x \wedge y) - F(x)F(y)] dx dy$$

We can even obtain an asymptotically efficient L-estimator with a suitable choice of the weight function; more precisely, the efficient weight function has the form

$$J(u) = J_F(u) = \frac{\psi'(F^{-1}(u))}{\mathcal{I}(F)}, \quad 0 < u < 1$$

$$\text{where} \qquad \psi(x) = -\frac{f'(x)}{f(x)}, \quad x \in \mathbb{R} \tag{6.10}$$

and $\mathcal{I}(F) = \int [f'(x)/f(x)]^2 dF(x)$ is the Fisher information of F; this has a sense only if $0 < \mathcal{I}(F) < \infty$. The asymptotic variance of the resulting efficient L-estimator is

$$\sigma^2(J, F) = \frac{1}{\mathcal{I}(F)},$$

that is the Rao-Cramér lower bound. Notice that, in the case of $J_F(u) = 0$ for $0 < u < \alpha$ and for $1 - \alpha < u < 1$, then $\frac{f'(x)}{f(x)} = \frac{d}{dx} \log f(x) = \text{const}$ for $x < F^{-1}(\alpha)$ and $x > F^{-1}(1 - \alpha)$, and hence the tails of the density $f(x)$ should decrease to 0 as fast as e^{-x} or e^x as $x \to \pm\infty$.

Let us now turn to the L-estimators of type II, that are linear combinations of a finite number of quantiles, $T_n = \sum_{i=1}^{k} a_j X_{n:[np_j]+1}$, $0 < p_1 < \ldots < p_k < 1$. The asymptotic behavior of such estimators is supported if the quantile function F^{-1} is smooth in neighborhoods of $p_1, \ldots, p_k$, e.g., if F is twice differentiable at $F^{-1}(p_j)$ and $F'\left(F^{-1}(p_j)\right) > 0$, $j = 1, \ldots, k$. Then, as $n \to \infty$,

$$\sqrt{n}\left(T_n - \sum_{j=1}^{k} a_j F^{-1}(p_j)\right) = n^{-1/2} \sum_{i=1}^{n} \psi_2(X_i) + R_n$$

$$\text{where} \qquad R_n = \mathcal{O}\left(n^{-1/4}(\log n)^{1/2}(\log\log n)^{1/4}\right) \text{ a.s.} \tag{6.11}$$

$$\text{and} \qquad \psi_2(x) = \sum_{j=1}^{k} \frac{a_j}{F'\left(F^{-1}(p_j)\right)} \{p_j - I\left[x \leq F^{-1}(p_j)\right]\}, \quad x \in \mathbb{R}$$

Then $\sqrt{n}\left(T_n - \sum_{j=1}^{k} a_j F^{-1}(p_j)\right)$ is asymptotically normally distributed with the law $\mathcal{N}\left(0, \int_{\mathbb{R}} \psi_2^2(x) dF(x)\right)$.

6.4 R-estimators

We shall illustrate the special case of estimating the center of symmetry θ of distribution function $F(x-\theta)$ with the aid of an R-estimator T_n. It is generated by the rank statistic $S_n(t)$ (3.89) by means of relations (3.90). Assume that F has positive and finite Fisher information $\mathcal{I}(F)$ and that the score function $\varphi(u)$ in (3.89) is nondecreasing and square-integrable, $0 < u < 1$. Then the asymptotic representation of T_n has the form

$$\sqrt{n}(T_n - \theta) = \frac{1}{\sqrt{n}\gamma}\sum_{i=1}^{n}\varphi(F(X_i - \theta)) + o_p(1) \quad \text{as} \ \ n \to \infty \qquad (6.12)$$

where $\gamma = \int_{\mathbb{R}}\varphi(F(x))(-f'(x))dx$. This in turn implies that $\sqrt{n}(T_n - \theta)$ has an asymptotically normal distribution $\mathcal{N}\left(0, \gamma^{-2}\int_0^1 \varphi^2(u)du\right)$. We even obtain an asymptotically efficient R-estimator with the asymptotic variance $1/\mathcal{I}(F)$ if we take

$$\varphi(u) = -\frac{f'(F^{-1}(u))}{f(F^{-1}(u))}, \quad 0 < u < 1.$$

6.5   Interrelationships of M-, L- and R-estimators

Consider two sequences of estimators $\{T_{1n}\}$ and $\{T_{2n}\}$ of θ, which is the center of symmetry of distribution $F(x - \theta)$. Assume that both $\{T_{1n}\}$ and $\{T_{2n}\}$ are asymptotically normally distributed, namely that $\sqrt{n}(T_{jn} - \theta)$ has asymptotic normal distribution $\mathcal{N}(0, \sigma_j^2)$, $j = 1, 2$, as $n \to \infty$. Then the ratio of variances $e_{1,2} = \sigma_1^2/\sigma_2^2$ is called the *asymptotic relative efficiency* of $\{T_{2n}\}$ with respect to $\{T_{1n}\}$. Alternatively, if $\{T_{2n'}\}$ is based on n' observations, then both $\sqrt{n}(T_{2n'} - \theta)$ and $\sqrt{n}(T_{1n} - \theta)$ have asymptotically normal distribution $\mathcal{N}(0, \sigma_1^2)$, provided the sequence $n' = n'(n)$ is such that there exists the limit

$$\lim_{n \to \infty}\frac{n}{n'(n)} = \frac{\sigma_1^2}{\sigma_2^2} = e_{1,2}$$

The fact that $e_{1,2} = 1$ means that $\{T_{1n}\}$ and $\{T_{2n}\}$ are equally asymptotically efficient. If this is the case, we can consider a finer comparison of $\{T_{1n}\}$ and $\{T_{2n}\}$ by means of the so-called *deficiency* of $\{T_{2n}\}$ with respect to $\{T_{1n}\}$. If the second moments of T_{n1} and T_{n2} have approximations

$$\mathrm{E}_\theta\left[n(T_{nj} - \theta)^2\right] = \tau^2 + \frac{a_j}{n} + o(n^{-1}), \ j = 1, 2$$

then the ratio

$$d_{1,2} = \frac{a_2 - a_1}{\tau^2}$$

is called the *deficiency* of $\{T_{2n}\}$ with respect to $\{T_{1n}\}$. If $n'(n)$ is chosen so that

$$\mathrm{E}_\theta[n(T_{2n'} - \theta)^2] = \mathrm{E}_\theta[n(T_{1n} - \theta)^2] + \mathcal{O}(n^{-1})$$

then

$$d_{1,2} = \lim_{n \to \infty} [n'(n) - n]$$

If $J(u) = \psi'(F^{-1}(u))$, $0 < u < 1$, then the influence functions of the M-estimator and L-estimator based on observations with distribution function $F(x - \theta)$ coincide, i.e., $IF(x, T_1, F) \equiv IF(x, T_2, F)$. This was already mentioned in Chapter 3. Similar relations hold also between M-estimators and R-estimators, and between L-estimators and R-estimators. Using the asymptotic representations (6.4), (6.9), (6.11), (6.12), we can claim even stronger statements. Comparing these representations, we see that if they coincide (up to the remainder terms), then the estimators not only have the same influence functions, but the sequences $\{T_{n1}\}$ and $\{T_{n2}\}$ are asymptotically close to each other in the sense that

$$\sqrt{n}(T_{2n} - T_{1n}) = R_n = o_p(1) \quad \text{as} \quad n \to \infty. \qquad (6.13)$$

Then we say that sequences $\{T_{n1}\}$ and $\{T_{n2}\}$ of estimators are *asymptotically equivalent*. If we are able to derive the exact order of the remainder term R_n in (6.13), we can get more precise information on the relation of $\{T_{n1}\}$ and $\{T_{n2}\}$. Sometimes we are able to find the asymptotic distribution of R_n, standardized by an appropriate power of n (most often $n^{\frac{1}{2}}$ or $n^{\frac{1}{4}}$). This is an asymptotic distribution of the second order and it is not normal.

Unfortunately, the conditions for (6.13) depend on the unknown distribution function F, thus we cannot derive the value of the R-estimator, once we have calculated the M-estimator, and so on. These relationships have rather a theoretical character, but they enable us to discover properties of one class of estimators, once we have proven it for another class.

Let us summarize the most interesting asymptotic relations among estimators of different types.

6.5.1   M- and L-estimators

Let $X_1, X_2, \ldots$ be independent random variables with a joint distribution function $F(x - \theta)$ such that $F(x) + F(-x) = 1$, $x \in \mathbb{R}$; let $X_{n:1} \leq X_{n:2} \leq \ldots \leq X_{n:n}$ be the order statistics corresponding to $X_1, \ldots, X_n$.

I. Let M_n be the M-estimator of θ generated by a nondecreasing step function ψ,

$$\psi(x) = \alpha_j \quad \ldots \quad s_j < x < s_{j+1}, \ j = 1, \ldots, k \qquad (6.14)$$

where

$$-\infty = s_0 < s_1 < \ldots < s_k < s_{k+1} = \infty$$

$$-\infty < \alpha_0 \leq \alpha_1 \leq \ldots \leq \alpha_k < \infty$$

$$\alpha_j = -\alpha_{k-j+1}, \ s_j = -s_{k-j+1}, \ j = 1, \ldots, k$$

and at least two among $\alpha_1, \ldots, \alpha_k$ are different. It means that M_n is a solution of the minimization $\sum_{i=1}^{n} \rho(X_i - t) = \min$, where ρ is a continuous, convex, symmetric and piecewise linear function with derivative $\rho' = \psi$ a.e. Assume that F has two bounded derivatives f, f' and that f is positive in neighborhoods of $s_1, \ldots, s_k$.

Then the L-estimator L_n, asymptotically equivalent to M_n, is the linear combination of a finite number of quantiles, $L_n = \sum_{j=1}^{k} a_j X_{n:[np_j]}$, where

$$p_j = F(s_j), \quad a_j = \frac{1}{\gamma}(\alpha_j - \alpha_{j-1}) f(s_j)$$

$$\gamma = \sum_{j=1}^{k} (\alpha_j - \alpha_{j-1}) f(s_j) \ (> 0) \tag{6.15}$$

$$\text{and } M_n - L_n = \mathcal{O}_p\left(n^{-\frac{3}{4}}\right) \quad \text{as } n \to \infty$$

II. Assume that F has an absolutely continuous symmetric density f and finite Fisher information $\mathcal{I}(F) > 0$. Let M_n be the Huber M-estimator of θ, generated by the function ψ

$$\psi(x) = \begin{cases} x & \dots & |x| \le c \\ c \cdot \text{sign } x & \dots & |x| > c \end{cases}$$

where $c > 0$. Let L_n be the α-trimmed mean,

$$L_n = \frac{1}{n - 2[n\alpha]} \sum_{i=[n\alpha]+1}^{n-[n\alpha]} X_{n:i}$$

where $\alpha = 1 - F(c)$. If F further satisfies $f(x) > a > 0$ and $f'(x)$ exists for

$$x \in \left(F^{-1}(\alpha - \varepsilon), F^{-1}(1 - \alpha + \varepsilon)\right) \quad \varepsilon > 0,$$

then

$$M_n - L_n = \mathcal{O}_p\left(n^{-1}\right) \tag{6.16}$$

as $n \to \infty$.

III. Let L_n be the α-Winsorized mean

$$L_n = \frac{1}{n}\left\{ [n\alpha]X_{n:[n\alpha]+1} + \sum_{i=[n\alpha]+1}^{n-[n\alpha]} X_{n:i} + [n\alpha]X_{n:n-[n\alpha]} \right\}$$

Then, under the same conditions as in **II**,

$$M_n - L_n = \mathcal{O}_p\left(n^{-\frac{3}{4}}\right), \quad n \to \infty \tag{6.17}$$

where M_n is the M-estimator generated by the function

$$
\psi(x) = \begin{cases}
F^{-1}(\alpha) - \frac{\alpha}{f(F^{-1}(\alpha))} & x < F^{-1}((\alpha) \\
x & F^{-1}(\alpha) \leq x \leq F^{-1}(1-\alpha) \\
F^{-1}(1-\alpha) + \frac{\alpha}{f(F^{-1}(\alpha))} & x > F^{-1}(1-\alpha)
\end{cases}
$$

IV. Let $L_n = \sum_{i=1}^{n} c_{ni} X_{n:i}$, where the coefficients c_{ni} are generated by the function $J : (0, 1) \mapsto \mathbb{R}$ such that

$$
J(1-u) = J(u), \ 0 < u < 1, \quad \int_0^1 J(u)du = 1
$$

$$
J(u) = 0 \ \text{ for } \ u \in (0, \alpha) \cup (1-\alpha, 1), \ 0 < \alpha < \frac{1}{2}
$$

and such that J is continuous in $(0, 1)$ up to a finite number of points $s_1, \ldots, s_m$, where $\alpha < s_1 \ldots < s_m < 1 - \alpha$, and J is Lipschitz continuous in intervals $(\alpha, s_1), (s_1, s_2), \ldots, (s_m, 1 - \alpha)$. The distribution function F is assumed to have a symmetric density, and $F^{-1}(u) = \inf\{x : F(x) \geq u\}$ is supposed to be Lipschitz continuous in the neighborhood of $s_1, \ldots, s_m$, and

$$
\int_{-A}^{A} f^2(x)dx < \infty, \quad \text{where} \ \ A = F^{-1}(1 - \alpha + \varepsilon), \ \varepsilon > 0
$$

Then an asymptotically equivalent M-estimator M_n is generated by the function

$$
\psi(x) = -\int_{\mathbb{R}} (I[y \geq x] - F(y)) J(F(y))dy, \ x \in \mathbb{R}
$$

and

$$
M_n - L_n = \mathcal{O}_p\left(n^{-1}\right), \ n \to \infty. \tag{6.18}
$$

6.5.2 *M- and R-estimators*

Again, let $X_1, X_2, \ldots$ be independent random variables with the same distribution function $F(x - \theta)$ such that $F(x) + F(-x) = 1$, $x \in \mathbb{R}$. We assume that F has an absolutely continuous density f and finite Fisher information $\mathcal{I}(F) > 0$. Let $\varphi : (0, 1) \mapsto \mathbb{R}$ be a nondecreasing score function, $\varphi(1 - u) = -\varphi(u)$, $0 < u < 1$ and $\int_0^1 \varphi^2(u)du < \infty$. Assume that

$$
\gamma = -\int_0^1 \varphi(F(x))f'(x)dx > 0
$$

Let R_n be the R-estimator, defined in (3.89) and (3.90) with the scores $a_n(i) = \varphi^+\left(\frac{i}{n+1}\right)$, $i = 1, \ldots, n$, where $\varphi^+(u) = \varphi\left(\frac{u+1}{2}\right)$, $0 \leq u < 1$. Moreover, let M_n be the M-estimator $\psi(x) = c\varphi(F(x))$, $x \in \mathbb{R}$, $c > 0$. Then

$$
\sqrt{n}(M_n - R_n) = o_p(1) \ \text{ as } \ n \to \infty. \tag{6.19}
$$

Particularly, the Hodges-Lehmann R-estimator is generated by the score function $\varphi(u) = u - \frac{1}{2}$, $0 < u < 1$. Hence, an asymptotically equivalent M-estimator is generated by the ψ-function $\psi(x) = F(x) - \frac{1}{2}$, $x \in \mathbb{R}$.

6.5.3 R- and L-estimators

Combining the previous results, we obtain the asymptotic relations of R- and L-estimators. An interesting special case is the R-estimator, asymptotically equivalent to the α-trimmed mean. It is generated by the score function

$$\varphi(u) = \begin{cases} F^{-1}(\alpha) & \dots \quad 0 < u < \alpha \\ F^{-1}(u) & \dots \quad \alpha \leq u \leq 1 - \alpha \\ F^{-1}(1 - \alpha) & \dots \quad 1 - \alpha < u < 1. \end{cases}$$

6.6 Estimation under contaminated distribution

Many estimators are asymptotically normally distributed, i.e., the distribution of $\sqrt{n}(T_n - T(F))$ converges to the normal distribution $\mathcal{N}(0, V_{as}(F, T))$ as $n \to \infty$, where $V_{as}(F, T) = \int_{\mathbb{R}} IF^2(x, T, F) dF(x)$.

The maximum asymptotic variance

$$\sigma^2(T) = \sup_{F \in \mathcal{F}} V_{as}(F, T)$$

over a specified class $\mathcal{F}$ of distribution functions can be considered as a measure of robustness of the functional T (and of its estimator T_n). On the other hand, consider a class of functionals $\mathcal{T}$ (e.g., of M-functionals), and look for $T_0 \in \mathcal{T}$ that $\sigma^2(T_0) \leq \sigma^2(T)$, $\forall T \in \mathcal{T}$. If such a functional exists, then we call it a *minimaximally robust estimator* of $\mathcal{T}$; it satisfies

$$\sigma^2(T_0) = \inf_{T \in \mathcal{T}} \sup_{F \in \mathcal{F}} V_{as}(F, T). \tag{6.20}$$

Let us illustrate this situation on the special case of the location parameter, when $X_1, \dots, X_n$ is a random sample from a population with distribution function $F(x - \theta)$, θ is an unknown parameter, and F is an unknown element of a family $\mathcal{F}$ of distribution functions. Specifically, consider the following classes $\mathcal{F}$:

(i) *Contamination model:*

$$\mathcal{F}_G = \{F : \ F = (1 - \varepsilon)G + \varepsilon H, \ H \in \mathcal{P}\} \tag{6.21}$$

where G is a fixed distribution function, $\varepsilon \in [0, 1)$ is a fixed number, and H runs over a fixed class $\mathcal{P}$ of distribution functions.

(ii) *Kolmogorov model:*

$$\mathcal{F}_G = \left\{F : \ \sup_{x \in \mathbb{R}} |F(x) - G(x)| \leq \varepsilon\right\}, \ \varepsilon \in [0, 1) \ \text{fixed.} \tag{6.22}$$

Let $F_0 \in \mathcal{F}$ be the distribution function with the smallest Fisher information in $\mathcal{F}$ (the least favorable distribution of the family $\mathcal{F}$), i.e.,

$$\mathcal{I}(F_0) = \int_{\mathbb{R}} \left(\frac{f_0'(x)}{f_0(x)} \right)^2 dF_0 = \min_{F \in \mathcal{F}} \mathcal{I}(F).$$

Let $T_0 \in \mathcal{T}$ be the estimator, which is an asymptotically efficient estimator of θ for distribution function F_0, i.e., $V_{as}(F_0, T_0) = \frac{1}{\mathcal{I}(F_0)}$ (if it exists). If

$$\frac{1}{\mathcal{I}(F_0)} = V_{as}(F_0, T_0) \geq \sup_{F \in \mathcal{F}} V_{as}(F, T_0),$$

then

$$\inf_{T \in \mathcal{T}} \sup_{F \in \mathcal{F}} V_{as}(F, T) = \frac{1}{\mathcal{I}(F_0)}$$

i.e.,

$$\forall T \in \mathcal{T} \text{ and } \forall F \in \mathcal{F} \quad V_{as}(F_0, T) \geq V_{as}(F_0, T_0) \geq V_{as}(F, T_0). \qquad (6.23)$$

In the symmetric contamination model, the minimaximally robust estimators really exist in the classes of M-, L- and R-estimators, respectively (see Huber (1964), Jaeckel (1971) for more detail).

6.6.1 Minimaximally robust M-, L- and R-estimators

Consider the contamination model (6.21), where G is a symmetric *unimodal* distribution function with twice differentiable density g such that $(-\log g(x))$ is convex in x; let H run over symmetric distribution functions. The family of such contaminated distribution functions we denote $\mathcal{F}_1$. Let $T(F)$ be the M-functional, defined as a root of the equation $\int_{\mathbb{R}} \psi(x - T(F)) dF(x) = 0$. Then

$$V_{as}(F, T) = \frac{\int_{\mathbb{R}} \psi^2(x - T(F)) dF(x)}{\left(\int_{\mathbb{R}} \psi'(x - T(F)) dF(x) \right)^2} \geq \frac{1}{\mathcal{I}(F)}$$

The least favorable distribution of the family $\mathcal{F}_1$ has the density (see Huber (1964))

$$f_0(x) = \begin{cases} (1 - \varepsilon)g(x_0)e^{k(x - x_0)} & \cdots & x \leq x_0 \\ (1 - \varepsilon)g(x) & \cdots & x_0 \leq x \leq x_1 \\ (1 - \varepsilon)g(x_1)e^{-k(x - x_1)} & \cdots & x \geq x_1 \end{cases} \qquad (6.24)$$

where

$$x_0 = -x_1 = \inf \left\{ x : \ -\frac{g'(x)}{g(x)} \geq -k \right\}$$

and $k > 0$ is determined by the relation

$$\frac{2}{k}g(x_1) + \int_{x_0}^{x_1} g(x)dx = \frac{1}{1 - \varepsilon}$$

T_n is the maximal likelihood estimator for distribution f_0, hence the M-estimator generated by the function

$$\psi_0(x) = -\frac{f_0'(x)}{f_0(x)} = \begin{cases} -k & \dots & x \leq x_0 \\ -\frac{g'(x)}{g(x)} & \dots & x_0 < x < x_1 \\ k & \dots & x \geq x_1 \end{cases}$$

We see from the asymptotic relations of Section 6.5 that minimaximally robust estimators also exist in the classes of L- and R- estimators. Particularly, the minimaximally robust L-estimator is generated by the weight function

$$J_0(u) = \frac{1}{\mathcal{I}(F_0)} \psi_0'(F^{-1}(u)), \ 0 < u < 1$$

and the minimaximally robust R-estimator is generated by the score function

$$\varphi_0(u) = \psi_0(F_0^{-1}(u)), \ 0 < u < 1.$$

An important special case is the minimaximally robust estimator in the model of the *contaminated normal distributions:* More precisely, insert the $\mathcal{N}(0,1)$ distribution function as $G \equiv \Phi$ into the model (6.21). Then the corresponding least favorable distribution has the density

$$f_0(x) = \begin{cases} \frac{1-\varepsilon}{\sqrt{2\pi}} e^{-x^2/2} & \dots & |x| \leq k \\ \frac{1-\varepsilon}{\sqrt{2\pi}} e^{-k^2/2 - k|x|} & \dots & |x| > k. \end{cases} \tag{6.25}$$

Notice that it is normal in the central part $[-k, k]$ and exponential outside. The likelihood function of f_0 is

$$\psi_0(x) = -\frac{f_0'(x)}{f_0(x)} = \begin{cases} x & \dots & |x| \leq k \\ k \operatorname{sign} x & \dots & |x| > k. \end{cases}$$

It is the well-known Huber function. The constant $k > 0$ is determined by the relation

$$2\Phi(k) - 1 + 2\frac{\Phi'(k)}{k} = \frac{1}{1-\varepsilon}.$$

The minimaximally robust M-estimator for the contaminated normal distribution is generated by the function ψ_0; it coincides with the maximal likelihood estimator corresponding to the density f_0. The minimaximally robust L-estimator is generated by the weight function J_0 satisfying

$$J_0(F_0(x)) = \frac{1}{\mathcal{I}(F_0)} I[-k \leq x \leq k], \ x \in \mathbb{R}.$$

Hence, it equals

$$J_0(u) = \frac{1}{\mathcal{I}(F_0)} I\left[F_0^{-1}(-k) \leq u \leq F_0^{-1}(k)\right], \ 0 < u < 1.$$

The corresponding L-estimator is the α-trimmed mean, where $\alpha = F_0^{-1}(-k)$. Similarly, the minimaximally robust R-estimator for the contaminated normal distribution is generated by the score function $\varphi_0(u) = \psi_0\left(F_0^{-1}(u)\right)$, $0 < u < 1$.

6.7   Possible non-admissibility under finite-sample

The asymptotic (normal) distribution of an estimator approximates well the central part of its true distribution, but not so much its tails. Robust estimators, advertised as resistant to heavy-tailed distributions, can be themselves heavy-tailed, though asymptotically normal. Even when asymptotically admissible, many are not finite-sample admissible under any distribution extended over the whole real line. Hence, before taking a recourse to the asymptotics, we should first analyze finite-sample properties of an estimator, whenever possible. We shall illustrate some distinctive differences between the asymptotic and finite-sample admissibility of robust estimators.

It turns out that the location estimators, which trim-off the extreme order statistics of the sample, cannot be admissible, not even Bayesian in the class of translation equivariant estimators, whatever the strictly convex and continuously differentiable loss function. This phenomenon extends to the linear regression model; this among others implies the inadmissibility of the trimmed LSE of Koenker and Bassett (1978) with respect to L_p ($p \geq 2$) or to other smooth convex loss functions.

6.7.1   Trimmed estimator cannot be Bayesian

Let $X_1, \ldots, X_n$ be a random sample from a population with density $f(x, \theta) > 0$, $\theta \in \Theta$, where Θ is an open subset of real line $\mathbb{R}^1$; let $X_{n:1} \leq X_{n:2} \leq \ldots \leq X_{n:n}$ be the order statistics. The loss incurred when estimating θ by $t \in \mathbb{R}$ is $L(t - \theta) \geq 0$, continuously differentiable and strictly convex. Let Λ be a prior distribution on parametric space Θ; we assume that L is square integrable with respect to the measures $f(x, \theta)d\Lambda(\theta)$, $x \in \mathbb{R}$. Note that Λ can be a generalized distribution, i.e., Λ needs not to be a probability measure; it can be e.g., a σ-finite measure.

Let $\theta^* = \theta^*(X_1, \ldots, X_n)$ be the generalized Bayesian estimator of θ with respect to the prior Λ, i.e., the solution of the equation

$$\int_\Theta L'(\theta^* - \theta) f(X_1, \theta) \cdots f(X_n, \theta) d\Lambda(\theta) = 0. \tag{6.26}$$

Our question is whether the generalized Bayes estimator θ^* can be independent of both extremes $X_{1:n}$ and $X_{n:n}$. The answer to this question is negative if $f(x, \theta)$ is positive and continuous in $x \in \mathbb{R}$, $\theta \in \Theta$ and the linear space over the functions $f(x, .)$ is dense in the space of all functions of θ, integrable with respect to the measure Λ. The measure Λ should have continuous density [not necessarily probabilistic], and the loss function $L(t - \theta) \geq 0$ should be continuously differentiable, strictly convex and square integrable with respect to the measures $f(x, \theta)d\Lambda(\theta)$, $x \in \mathbb{R}$. Then we have the following theorem:

Theorem 6.1 *Under the above assumptions, the estimator θ^* functionally depends at least on one of the extreme observations $X_{n:1}, X_{n:n}$.*

PROOF. Under the assumptions, the equation

$$\int_\Theta L'(\theta^* - \theta)f(X_{1:n}, \theta) \cdots f(X_{n:n}, \theta)d\Lambda(\theta) = 0 \qquad (6.27)$$

has the only solution θ^*. By Lehmann (1997), Theorem 4.3.1, such θ^* is admissible. Assume that θ^* depends neither on $X_{1:n}$ nor on $X_{n:n}$. Then we obtain

$$\int_\Theta L'(\theta^*(X_{2:n}, \ldots, X_{(n-1):n}) - \theta)f(X_{1:n}, \theta) \cdots f(X_{n:n}, \theta)d\Lambda(\theta) = 0. \quad (6.28)$$

Put $X_{1:n} = t$, $X_{2:n} = \ldots = X_{(n-1):n} = s$, $X_{n:n} = s$ for a fixed s and arbitrary $t \leq s$. Denoting $\theta_s = \theta^*(s, \ldots, s)$, we can rewrite (6.28) in the form

$$\int_\Theta L'(\theta_s - \theta)f(t, \theta)f^{n-1}(s, \theta)d\Lambda(\theta) = 0 \qquad (6.29)$$

for all $t \leq s$. Analogously, the relation (6.29) is valid for all $t \geq s$, when we put $X_{1:n} = X_{2:n} = \ldots = X_{(n-1):n} = s$, $X_{1:n} = t$, $t \geq s$ in (6.28). Hence, the relation (6.29) is true for any fixed s and all t. Because the linear space over the functions $f(x, \theta)$ is dense among all functions of θ integrable with respect to the measure Λ, then

$$L'(\theta_s - \theta)f^{n-1}(s, \theta) = 0 \text{ a.s. } [\Lambda],$$

which is in contradiction with the assumption that f is positive continuous density and L is strictly convex. □

Particularly, consider the minimum risk equivariant estimator T_n of location parameter θ with respect to the L^2 norm. It is the *Pitman estimator* which can be expressed as

$$T_n = \hat{\theta}_n - E_0(\hat{\theta}_n | \mathbf{Y}) \qquad (6.30)$$

where $\hat{\theta}_n$ is any equivariant estimator with a finite risk and $\mathbf{Y}$ is a maximal invariant with respect to the group of translations, for instance

$$\mathbf{Y} = (X_2 - X_1, \ldots, X_n - X_1) \quad \text{or} \quad \mathbf{Y} = (X_1 - \bar{X}, \ldots, X_n - \bar{X}). \quad (6.31)$$

Notice that T_n is the generalized Bayesian estimator with respect to the uniform prior with the density $\pi(\theta) = \frac{d\Lambda(\theta)}{d\theta} = 1$, $\theta \in \mathbf{R}^1$.

On the other hand, every equivariant estimator $\hat{\theta}_n$ with finite quadratic risk satisfies the inequalities

$$E_\theta(\hat{\theta}_n - \theta - E_0(\hat{\theta}_n | \mathbf{Y}))^2 = E_\theta(\hat{\theta}_n - \theta)^2 - E_0(E_0(\hat{\theta}_n | \mathbf{Y}))^2 \leq E_\theta(\hat{\theta}_n - \theta)^2. \quad (6.32)$$

Hence, an equivariant estimator $\hat{\theta}_n$ has the minimal risk if and only if $E_0(\hat{\theta}_n|\mathbf{Y}) = 0$. We get a consequence of Lemma 6.1:

Corollary 6.1 *Let $X_1, \ldots, X_n$ be independent observations with continuous density $f(x - \theta) > 0$, $\theta \in \mathbb{R}$, satisfying the conditions of Theorem 6.1.*

(i) *Then the Pitman estimator functionally depends either on $X_{n:1}$ or on $X_{n:n}$. Consequently, no admissible estimator can be independent of both $X_{n:1}$ and $X_{n:n}$.*

(ii) *If T_n is a robust estimator, trimming some smallest or largest observations, then it is inadmissible in quadratic loss function for any distribution satisfying the conditions of Theorem 6.1.*

This illustrates a possible inadmissibility of some robust estimators. It is true in a much wider context: A similar statement is true for other convex loss functions and for more general models, as the linear one.

6.8   Newton-Raphson iterations of estimating equations

Some estimators, such as the M-estimators, regression quantiles and maximal likelihood estimators, are defined implicitly as solutions of a minimization or of a system of equations. It can be difficult to solve such problems algebraically: The system of equations can have more solutions, with only one being efficient, and it is hard to distinguish the efficient solution from the others, if it is possible at all. We have already met this situation in the context of M-estimators, where the system of Equations (4.30) can have more roots, among them at least one is $\sqrt{n}$-consistent, but it is hard to distinguish it from the others.

The $\sqrt{n}$-consistent root of the system of equations, including the efficient one, can often be approximated by its one-step version. This consists of the first step of the Newton-Raphson iteration algorithm of solving the algebraic equation. This approach was first used by Bickel (1975) for the one-step version of the Huber M-estimator in the linear regression model (4.1).

Generally, the iteration starts with an initial consistent estimator $\mathbf{M}_n^{(0)}$ of parameter $\boldsymbol{\beta}$; the $\sqrt{n}$-consistent initial estimator gives the best results.

Let us illustrate this approach on the one-step version of the M-estimator in the linear regression model (4.1), generated by a function ρ with derivative ψ, and studentized by scale statistics $S_n = S_n(\mathbf{Y})$. The one-step M-estimator is defined through the relation

$$\mathbf{M}_n^{(1)} = \begin{cases} \mathbf{M}_n^{(0)} + \frac{1}{\hat{\gamma}_n}\mathbf{W}_n & \cdots \quad \hat{\gamma}_n \neq 0 \\ \mathbf{M}_n^{(0)} & \cdots \quad \hat{\gamma}_n = 0 \end{cases} \tag{6.33}$$

where

$$\boldsymbol{W}_n = \boldsymbol{Q}_n^{-1} \sum_{i=1}^n \boldsymbol{x}_i \psi \left(\frac{Y_i - \boldsymbol{x}_i^\top \mathbf{M}_n^{(0)}}{S_n} \right)$$

$$\boldsymbol{Q}_n = \frac{1}{n} \boldsymbol{X}_n^\top \boldsymbol{X}_n$$

and $\hat{\gamma}_n$ is an estimator, either of the functional

$$\gamma = \frac{1}{S(F)} \int_{\mathbb{R}} \psi' \left(\frac{x}{S(F)} \right) dF(x)$$

or of the functional

$$\gamma = \int_{\mathbb{R}} f(xS(F)) d\psi(x)$$

depending on whether the selected score function ψ is continuous or not. For instance, in the case of absolutely continuous ψ, we can use the estimator

$$\hat{\gamma}_n = \frac{1}{nS_n} \sum_{i=1}^n \psi' \left(\frac{Y_i - \boldsymbol{x}_i^\top \mathbf{M}_n^{(0)}}{S_n} \right).$$

$\mathbf{M}_n^{(1)}$ is a good approximation of the consistent M-estimator $\mathbf{M}_n$: if $\mathbf{M}_n^{(0)}$ is $\sqrt{n}$-consistent and ψ is sufficiently smooth, then

$$\|\mathbf{M}_n - \mathbf{M}_n^{(1)}\| = \mathcal{O}_p \left(n^{-1} \right). \qquad (6.34)$$

However, if ψ has jump discontinuities, then

$$\|\mathbf{M}_n - \mathbf{M}_n^{(1)}\| = \mathcal{O}_p \left(n^{-3/4} \right) \quad \text{only.}$$

More on the one-step versions of M- and L-estimators can be found in Janssen et al. (1985), Portnoy and Jurečková (1999), and Jurečková and Welsh (1990); an intensive study of this approach was made by Linke (2013–2017), Linke and Borisov (2017), Linke and Sakhanenko (2013–2014), among others. The k-step versions in the location model were studied by Jurečková and Malý (1995).

Generally we can say that the one-step versions give good approximations of an M-estimator with smooth function ψ, while the approximation is rather poor for discontinuous ψ, and then it is not much improved even when we use k iterations instead of one.

6.8.1   Finite-sample and asymptotic properties of the iterations

The difference between various $\sqrt{n}$-consistent initial estimators is immediately apparent numerically, which demonstrates that the initial estimator plays a substantial role. The experience leads to a conjecture that the finite-sample properties of the one-step version (or even of the k-step version) of an M-estimator heavily depend on the properties of the initial estimator; this concerns not only the breakdown point, but also the tail-behavior measure (4.5) and others (see e.g., Mizera and Müller (1999), Jurečková (2012), Welsh and Ronchetti (2002)) compared various one-step versions of M-estimators, and effects of various initial estimators. The difference between various $\sqrt{n}$-consistent initial estimators appears only in the reminder terms of the approximations, or in the so-called second-order asymptotics. Surprisingly, though theoretically we always get an approximation (6.34), the numerical evidence shows that the initial estimator plays a substantial role. Asymptotically, the best approximation is given by an initial $\mathbf{M}_n^{(0)}$, which already has the same influence function as $\mathbf{M}_n$ (see Jurečková and Sen 1990). This is true for $\mathbf{M}_n^{(1)}$, as we see from (6.34), hence we should follow the Welsh and Ronchetti proposal and use two Newton-Raphson iterations, instead of one.

The one-step version of $\mathbf{M}_n$ was modified in Portnoy and Jurečková (1987), under some conditions on ψ and on F, in such a way that it inherited the breakdown point of the initial estimator $\mathbf{M}_n^{(0)}$ and the asymptotic efficiency of $\mathbf{M}_n$. More precisely, assume that F is symmetric and $\psi = \rho'$ skew-symmetric, absolutely continuous, non-decreasing and bounded. Define the one-step version in the following way:

$$
\mathbf{M}_n^{(1)} = \begin{cases} \mathbf{M}_n^{(0)} + \hat{\gamma}_n^{-1}\mathbf{W}_n & \cdots \quad \left\| \hat{\gamma}_n^{-1}\mathbf{W}_n \right\| \leq c,\ 0 < c < \infty \\ \mathbf{M}_n^{(0)} & \cdots \qquad \text{otherwise} \end{cases} \tag{6.35}
$$

$$
\hat{\gamma}_n = \frac{1}{2n^{1/2}} \sum_{i=1}^{n} x_{i1} \left[\frac{\psi(Y_i - \mathbf{x}_i^\top \mathbf{M}_n^{(0)} + n^{-1/2}\mathbf{x}_i^\top \mathbf{q}_n^{(1)})}{S_n} \right.
$$
$$
\left. - \frac{\psi(Y_i - \mathbf{x}_i^\top \mathbf{M}_n^{(0)} - n^{-1/2}\mathbf{x}_i^\top \mathbf{q}_n^{(1)})}{S_n} \right] \tag{6.36}
$$

where $\mathbf{q}_n^{(1)}$ is the first column of $\mathbf{Q}_n^{-1}$; some other smoothness conditions are assumed for brevity. Then

Theorem 6.2 *Under the above conditions,*

(i) $\left\| \mathbf{M}_n^{(1)} - \mathbf{M}_n \right\| = O_p(n^{-1}).$

(ii) *If $\mathbf{M}_n^{(0)}$ has a finite sample breakdown point m_n, then $\mathbf{M}_n^{(1)}$ has the same breakdown point m_n.*

(iii) *If $\mathbf{M}_n^{(0)}$ is affine equivariant, then $P\left\{ \mathbf{M}_n^{(1)}(\mathbf{X}_n\mathbf{A}) \neq \mathbf{A}^{-1}\mathbf{M}_n^{(1)}(\mathbf{X}_n) \right\} \to 0$ as $n \to \infty$ for any regular $p \times p$ matrix $\mathbf{A}$.*

See Portnoy and Jurečková (1987) for the proof.

Remark 6.1 *The iteration procedure with a high breakdown initial estimate was extended by Simpson et al. (1992) to the GM-estimators. The results* *are true even for the initial estimator satisfying* $\|\mathbf{M}_n^{(0)} - \boldsymbol{\beta}\| = O_p(n^{-\tau})$, $\frac{1}{4} < \tau \leq \frac{1}{2}$.

The finite-sample breakdown point, inherited from the initial estimator, remains constant even with an increasing number of iterations (Jurečková 2012).

6.9   Adaptive combination of estimation procedures

There exists a big variety of robust estimators in the linear model. Many of them enjoy excellent properties and are equipped with computational software. Then it is difficult for an applied statistician to decide which procedure to use. Motivations can be various: sometimes he/she can prefer simplicity to optimality and to other advantages and to use the classical least squares L_1-method or other reasonably simple method. One possible device in such situation is to combine two convenient estimation methods; in this way one can diminish the eventual shortages of both methods. This idea, simple as it is, was surprisingly not very much elaborated until recently. Taylor (1973) was probably the first who proposed combining the L_1 and the least squares methods. He was followed by Arthanari and Dodge (1981), who started with an estimation method based on a direct convex combination of the L_1 and LS methods; see also Dodge and Lindstrom (1981). Dodge (1984) extended this method to a convex combination of the L_1 and Huber's M-estimation methods.

Dodge and Jurečková (1987) observed that the convex combination of two methods can be adaptively optimized, based on the observations, so that the resulting estimator has the minimal asymptotic variance over all estimators of this kind and over a class of distributions indexed with a nuisance scale. This was further illustrated and extended by Dodge and Jurečková (1988, 1991), Dodge, Antoch and Jurečková (1991), among others.

The general schema of the adaptive combination procedure is as follows:

We consider a family of symmetric densities indexed by an appropriate scale measure s, with a unit density f_0:

$$\mathcal{F} = \left\{ f : f(z) = s^{-1} f_0\left(z/s\right), \ s > 0 \right\}. \tag{6.37}$$

The shape of f_0 is generally unknown; it only satisfies some regularity conditions. The unit element $f_0 \in \mathcal{F}$ has $s_0 = 1$. When we combine L_1 with another estimator, we take $s = 1/f(0)$. We assume throughout that the scale characteristic s is estimable by means of a consistent estimator s_n based on $Y_1, \ldots, Y_n$. Moreover, as a scale measure s_n should be regression-invariant and scale-equivariant, i.e.,

$$
\begin{array}{lll}
\text{(a)} & s_n(\mathbf{Y}) \overset{p}{\to} s & \text{as } n \to \infty \qquad\qquad (6.38) \\
\text{(b)} & s_n(\mathbf{Y} + \mathbf{Xb}) = s_n(\mathbf{Y}) \text{ for any } \mathbf{b} \in R^p & \text{(regression-invariance)} \\
\text{(c)} & s_n(c\mathbf{Y}) = c s_n(\mathbf{Y}) \text{ for } c > 0 & \text{(scale-equivariance)}.
\end{array}
$$

Our goal is to estimate the parameter $\boldsymbol{\beta}$ in the linear regression model $\mathbf{Y} = \mathbf{X}\boldsymbol{\beta} + \mathbf{U}$. We choose two criterion functions ρ_1, ρ_2, symmetric and convex, defining two respective estimators. Particularly, $\rho_1(z) = \mid z \mid$ and $\rho_2(z) = z^2$ if we want to combine LAD (i.e., L_1) and LS estimators. Put $\rho(z) = \delta\rho_1(z) + (1-\delta)\rho_2(z)$ with a specific δ, $0 \leq \delta \leq 1$, and define the estimator $\mathbf{T}_n(\delta)$ as a solution of the minimization problem

$$\sum_{i=1}^{n} \rho\left(\frac{Y_i - \mathbf{x}_i^\top \mathbf{b}}{s_n}\right) := \min, \ \mathbf{b} \in \mathbb{R}^p. \tag{6.39}$$

Then $\sqrt{n}(\mathbf{T}_n(\delta) - \boldsymbol{\beta})$ typically has an asymptotically normal distribution $\mathcal{N}_p(\mathbf{0}, \mathbf{Q}^{-1}\sigma^2)$, where $\sigma^2 = \sigma^2(\delta, \rho, f)$ and $\mathbf{Q} = \lim_{n\to\infty} n^{-1}(\mathbf{X}_n^\top \mathbf{X}_n)$. Using $\delta = \delta_0$ which minimizes $\sigma^2(\delta, \rho, f)$ with respect to δ, $0 \leq \delta \leq 1$, we obtain an estimator $\mathbf{T}_n(\delta_0)$ minimizing the asymptotic variance over a family (6.37) of densities of a fixed distribution shape, differing only in the scale.

We shall illustrate in detail the adaptive combination of two estimation procedures on the combination of the most usual procedures, the L_1 and L_2 (the LAD and the LSE).

6.9.1 Adaptive convex combination of LAD and LS regression

The convex combination of L_1 and L_2 criteria has the form

$$(1 - \delta) \|\mathbf{Y} - \mathbf{X}\beta\|_2 + \delta \|\mathbf{Y} - \mathbf{X}\beta\|_1, \ 0 \leq \delta \leq 1.$$

When $\delta = 0$, we have the usual least squares method, which is the maximum likelihood estimator for the normal distribution. When $\delta = 1$, the minimization corresponds to the LAD-method, and the LAD-estimator is the maximal likelihood for the Laplace distribution. In this way, the adaptive procedure tries to decide how much the distribution of our observations leans to the normal or to the Laplace distribution.

The minimization of the combined procedure leads to the studentized M-estimator $\mathbf{T}_n(\delta)$ of $\boldsymbol{\beta}$, which is the solution of the minimization

$$\sum_{i=1}^{n} \rho((Y_i - \mathbf{x}_i^\top \mathbf{b})/s_n) := \min \tag{6.40}$$

with respect to $\mathbf{b} \in \mathbb{R}^p$, where $\mathbf{x}_i^\top$ is the i-th row of $\mathbf{X}_n$, $s_n = s_n(Y_1, \ldots, Y_n)$ is a suitable estimator of s and

$$\rho(z) = (1 - \delta)z^2 + \delta \mid z \mid, z \in \mathbb{R}^1 \tag{6.41}$$

with a fixed δ, $0 \leq \delta \leq 1$. The score function of $\mathbf{T}_n(\delta)$ is equal to

$$\psi(z) = 2(1 - \delta)z + \delta \ \mathrm{sign}\ z, \ z \in \mathbb{R}^1, \tag{6.42}$$

a nondecreasing function of z with a jump-discontinuity at $z = 0$, for every $\delta \in [0, 1]$. Set

$$\sigma^2 = \int z^2 f(z)dz, \qquad \sigma_0^2 = \int z^2 f_0(z)dz \qquad (6.43)$$

$$E_1^0 = \int |z| f_0(z)dz, \qquad E_1 = \int |z| f(z)dz.$$

Then

$$\sigma^2 = \int z^2 f(z)dz = s^2 \sigma_0^2 \qquad \text{and} \qquad E_1 = \int |z| f(z)dz = sE_1^0. \quad (6.44)$$

The estimator s_n of s should be consistent, regression-invariant and square equivariant; the estimators based on regression quantiles or on regression rank scores, described in Section 4.8, are good candidates for estimating s.

We shall consider the linear model with an intercept β_1; that means that design matrix $\mathbf{X}_n$ satisfies

$$x_{i1} = 1, \ i = 1, \dots, n \qquad (6.45)$$

The estimators under consideration will be asymptotically normally distributed under the following conditions:

$$\lim_{n \to \infty} \mathbf{Q}_n = \mathbf{Q} \qquad \text{where} \qquad \mathbf{Q}_n = n^{-1}\mathbf{X}_n^\top \mathbf{X}_n$$

and $\mathbf{Q}$ is a positively definite $(p \times p)$ matrix;

$$\max_{1 \leq i \leq n, 1 \leq j \leq p} |x_{ij}| = O(n^{1/4}) \qquad \text{as} \qquad n \to \infty \qquad (6.46)$$

$$n^{-1} \sum_{i=1}^{n} x_{ij}^4 = O(1) \qquad \text{as} \qquad n \to \infty \text{ for } j = 1, \dots, p.$$

The asymptotic distribution of the studentized M-estimator with the score function ψ has the form

$$\sqrt{n}(\mathbf{T}_n(\delta) - \boldsymbol{\beta}) \xrightarrow{\mathcal{D}} \mathcal{N}_p(\mathbf{0}, \mathbf{Q}^{-1}\sigma^2(\psi, F, s))$$

where

$$\sigma^2(\psi, F, s) = \int \psi^2(z/s)f(z)dz \left[\int f(sz)d\psi(z) \right]^{-2}$$

$$= \int \psi^2(z)f_0(z)dz \left[\int f(sz)d\psi(z) \right]^{-2} \qquad (6.47)$$

and this under special ψ of (6.42) leads to

$$\sigma^2(\psi, F, s) = \qquad (6.48)$$

$$= \frac{s^2}{4} \left\{ 4(1-\delta)^2 \int z^2 f_0(z)dz + \delta^2 \int f_0(z)dz + 4\delta(1-\delta) \int |z| f_0(z)dz \right\}$$

$$= \frac{s^2}{4} \left\{ 4(1-\delta)^2 \sigma_0^2 + \delta^2 + 4\delta(1-\delta)E_1^0 \right\}$$

Then δ_0, minimizing (6.48) with respect to $\delta \in [0,1]$, provides the convex combination of estimating procedures with the minimal asymptotic variance. More precisely,

$$\delta_0 = \begin{cases} 0 & \text{if } 2\sigma_0^2 \leq E_1^0 < 1/2 \\ \dfrac{4\sigma_0^2 - 2E_1^0}{4\sigma_0^2 - 4E_1^0 + 1} & \text{if } E_1^0 < 1/2 \text{ and } E_1^0 < 2\sigma_0^2 \\ 1 & \text{if } 1/2 \leq E_1^0 < 2\sigma_0^2. \end{cases} \qquad (6.49)$$

However, δ_0 depends on the moments of unknown f_0, thus they should be estimated with the aid of observations $Y_1, \ldots, Y_n$. We recommend the following procedure:

(i) Estimate s with $\hat{s}_n$ described in Section 4.8.

(ii) Estimate $\boldsymbol{\beta}$ with $\widehat{\boldsymbol{\beta}}_n = \widehat{\boldsymbol{\beta}}_n\left(\frac{1}{2}\right)$, the L_1-estimator of $\boldsymbol{\beta}$ (0.5-regression quantile).

(iii) Estimate $E_1^0 = E_1/s = f(0) \int | z | f(z)dz$ by

$$\widehat{E}_1^0 = \hat{s}_n^{-1}(n-p)^{-1} \sum_{i=1}^{n} | Y_i - \mathbf{x}_i' \widehat{\boldsymbol{\beta}}_n | . \qquad (6.50)$$

(iv) Estimate σ_0^2 with the estimator

$$\widehat{\sigma}_{0n}^2 = \frac{1}{\hat{s}_n^2(n-p)} \sum_{i=1}^{n} \left(Y_i - \mathbf{x}_i^\top \widehat{\boldsymbol{\beta}}_n \right)^2 . \qquad (6.51)$$

(v) If $\widehat{E}_{1n}^0 \geq 2\widehat{\sigma}_{0n}^2$, put $\widehat{\delta}_{0n} = 0$. Then $\mathbf{T}_n$ is the ordinary least squares estimator of $\boldsymbol{\beta}$.

(vi) If $\widehat{E}_{1n}^0 < 2\widehat{\sigma}_{0n}^2$, calculate

$$\widehat{\delta}_{0n} = \frac{4\widehat{\sigma}_{0n}^2 - 2\widehat{E}_{1n}^0}{4\widehat{\sigma}_{0n}^2 - 4\widehat{E}_{1n}^0 + 1}. \qquad (6.52)$$

Then $\mathbf{T}_n$ is the M-estimator minimizing (6.40) with ρ given in (6.41), more precisely, minimizing

$$(1 - \widehat{\delta}_{0n}) \sum_{i=1}^{n} \left(\frac{Y_i - \mathbf{x}_i^\top \mathbf{b}}{\hat{s}_n} \right)^2 + \widehat{\delta}_{0n} \sum_{i=1}^{n} \left| \frac{Y_i - \mathbf{x}_i^\top \mathbf{b}}{\hat{s}_n} \right| . \qquad (6.53)$$

(vii) If $\frac{1}{2} \leq \widehat{E}_{1n}^{(0)} < 2\widehat{\sigma}_{0n}^2$, put $\widehat{\delta}_{0n} = 1$. Then $\mathbf{T}_n$ is the LAD estimator of $\boldsymbol{\beta}$.

As it is shown in Dodge and Jurečková (1987), $\sqrt{n}(\mathbf{T}_n(\widehat{\delta}_{0n}) - \boldsymbol{\beta})$ is asymptotically normally distributed under the above conditions on $\mathbf{X}_n$ and f_0, i.e.,

$$\sqrt{n}(\mathbf{T}_n(\widehat{\delta}_{0n}) - \boldsymbol{\beta}) \xrightarrow{D} \mathcal{N}_p(\mathbf{0}, \sigma_0^2(\psi, F)\mathbf{Q}^{-1})$$

as $n \to \infty$, with $\sigma_0^2(\psi, F)$ minimizing (6.48), which is the minimum over $0 \le \delta \le 1$.

The procedure for adaptive convex combination of LAD and LS regression is not implemented in R yet. However, you can use the function LADLS, see Appendix A.

```
> library(quantreg)
> data(engel)
> LADLS (engel[,1], engel[,2], resultls, resultl1, int=T, kern=F)
 Intercept          X1
94.7966716  0.5490402


### LSE
> lm(engel[,2]~engel[,1])

Call:
lm(formula = engel[, 2] ~ engel[, 1])

Coefficients:
(Intercept)    engel[, 1]
   147.4754        0.4852


### LAD
> rq(engel[,2]~engel[,1], tau=0.5)
Call:
rq(formula = engel[, 2] ~ engel[, 1], tau = 0.5)

Coefficients:
(Intercept)    engel[, 1]
 81.4822474    0.5601806

Degrees of freedom: 235 total; 233 residual
```

The function LADLS was used also for the numerical illustration in the next subsection.

Table 6.1 *MSE in the sample of 5000 replications of selected estimators under different error distributions, sample size n = 25.*

Estimator	N(0,1)	N(0,4)	0.95 N(0,1) + 0.05 La(0,1)	0.90 N(0,1) + 0.10 La(0,1)	0.85 N(0,1) + 0.15 La(0,1)
LS	0.2509	3.6931	0.2451	0.2782	0.3502
LAD	0.3860	5.7527	0.3749	0.4226	0.4886
LMS	1.0675	15.5745	1.0745	1.1525	1.2767
M (Huber)	0.2654	3.9033	0.2597	0.2932	0.3540
L (TLS - KB)	0.2751	4.0554	0.2685	0.3066	0.3761
L (TLS - RC)	0.3012	4.4794	0.2949	0.3373	0.4110
LS-LAD	0.2521	3.6922	0.2455	0.2787	0.3477

	Lo(0,1)	Lo(0,4)	La(0,1)	La(0,4)	Ca(0,1)
LS	0.7770	12.243	0.4435	7.5331	548.178
LAD	0.9816	15.660	0.3754	6.1637	0.901
LMS	2.7477	46.182	1.0316	15.6334	1.640
M (Huber)	0.7396	11.445	0.3398	5.7970	1.141
L (TLS - KB)	0.8071	12.748	0.4320	7.3850	100.751
L (TLS - RC)	0.8446	13.502	0.4262	7.1308	3.495
LS-LAD	0.7665	12.132	0.4023	7.1414	2.067

6.10 Numerical illustration of LAD and LS regression

In order to illustrate how the adaptive convex combination of LAD and LS regression perform for finite sample situation we have conducted a simulation study.

We considered the linear regression model with the three artificial matrices of order 25×3, order 100×3, and order 500×3, respectively. The second and the third columns were prepared as a realization of the uniform distribution on $(-10,10)$ and with the first column of 1's and the "unknown" parameter $\beta = (5, -3, 1)'$. The underlying model errors were simulated from the following distributions: normal, logistic, Laplace, Cauchy, normal contaminated by Laplace. For each case, 5000 replications were simulated.

For a comparison the simulation study was made for several methods described in Chapter 4: the least squares (LS), the least absolute deviations (LAD), the least median of squares (LMS), Huber M (with c=1.2) estimators. Next, α-trimmed least squares (TLS - KB) studied by Koenker and Bassett and α-trimmed least squares (TLS - RC) estimator considered by

Table 6.2 *MSE in the sample of 5000 replications of selected estimators under different error distributions, sample sizes n = 100.*

Estimator	N(0,1)	0.95 N(0,1) + 0.05 La(0,1)	0.90 N(0,1) + 0.10 La(0,1)	0.85 N(0,1) + 0.15 La(0,1)
LS	0.0107	0.0093	0.0085	0.0079
LAD	0.0166	0.0148	0.0136	0.0126
LMS	0.0666	0.0646	0.0590	0.0536
M (Huber)	0.0113	0.0101	0.0092	0.0086
L (TLS - KB)	0.0112	0.0101	0.0093	0.0085
L (TLS - RC)	0.0113	0.0100	0.0092	0.0086
LS-LAD	0.0106	0.0093	0.0085	0.0079

	Lo(0,1)	La(0,1)	Ca(0,1)
LS	0.0329	0.0217	2237.2660
LAD	0.0442	0.0144	0.0300
LMS	0.1453	0.0423	0.0644
M (Huber)	0.0314	0.0158	0.0386
L (TLS - KB)	0.0318	0.0175	0.0714
L (TLS - RC)	0.0317	0.0170	0.0539
LS-LAD	0.0325	0.0170	0.0299

Ruppert and Carroll. Our main interest is to make out the performance of the adaptive convex combination of LAD and LS regression.

The selected results are summarized in Table 6.1 through Table 6.3, and the simulation study indicates that the LS-LAD estimator is quite stable under different distributions of the error terms and performs well even for moderate samples.

Table 6.3 *MSE in the sample of 5000 replications of selected estimators under different error distributions, sample sizes $n = 500$.*

Estimator	N(0,1)	0.95 N(0,1) + 0.05 La(0,1)	0.90 N(0,1) + 0.10 La(0,1)	0.85 N(0,1) + 0.15 La(0,1)
LS	0.00220	0.00199	0.00181	0.00167
LAD	0.00345	0.00312	0.00287	0.00266
M (Huber)	0.00238	0.00214	0.00195	0.00179
L (TLS - KB)	0.00234	0.00213	0.00194	0.00178
L (TLS - RC)	0.00236	0.00212	0.00193	0.00178
LS-LAD	0.00219	0.00199	0.00181	0.00167

	Lo(0,1)	La(0,1)	Ca(0,1)
LS	0.00694	0.00416	40464.57000
LAD	0.00804	0.00231	0.00506
M (Huber)	0.00626	0.00288	0.00862
L (TLS - KB)	0.00631	0.00311	0.01087
L (TLS - RC)	0.00631	0.00311	0.01020
LS-LAD	0.00679	0.00270	0.00506

6.11 Problems and complements

6.1 Let $X_1, \ldots, X_n$ have distribution function $F(x - \theta)$ where F is symmetric around 0. Let T_n be Huber's M-estimator generated by the ψ-function (3.16) with the boundary k. Then $\sqrt{n}(T_n - \theta)$ has asymptotically normal distribution $\mathcal{N}(0, \sigma^2(k, F))$ where

$$\sigma^2(k, F) = \frac{\int_{-k}^{k} x^2 dF(x) + 2k^2(1 - F(k))}{(F(k) - F(-k))^2}$$

6.2 Let $\mathcal{F}$ be the family of densities

$$\mathcal{F} = \left\{ f : f(x) = \frac{1}{s} f_0 \left(\frac{x}{s} \right), s > 0 \right\}$$

where f_0 is a fixed density (but generally unknown) such that

$$f_0(x) = f_0(-x), \quad x \in \mathbb{R}, \quad f_0(0) = 1, \quad \int x^2 f_0(x) dx < \infty$$

and f_0 has a bounded derivative in a neighborhood of 0.

Consider the linear regression model $Y_i = \mathbf{x}_i'\boldsymbol{\beta} + e_i$, $i = 1,\dots,n$ with $\boldsymbol{\beta} \in \mathbb{R}^p$, where $e_1,\dots,e_n$ are independent errors with joint density $f \in \mathcal{F}$, but unknown. The solution $\mathbf{T}_n(\delta)$ of the minimization

$$\sum_{i=1}^{n} \rho\left(\frac{Y_i - \mathbf{x}_i'\mathbf{b}}{\hat{s}_n}\right) = \min, \ \mathbf{b} \in \mathbb{R}^p$$

where $\rho(x) = (1-\delta)x^2 + \delta|x|$, $x \in \mathbb{R}$ and $\hat{s}_n$ is an estimator of s, is a M-estimator that can be considered as a mixture of the LSE and L_1 estimators of $\boldsymbol{\beta}$. For each $f \in \mathcal{F}$, there exists an optimal value $\delta_f \in [0,1]$, leading to the minimal asymptotic variance of $\mathbf{T}_n(\delta)$, $0 \le \delta \le 1$. The optimal δ_f can be estimated consistently by an estimator that depends only on $\hat{s}_n$ and on two first sample moments (Dodge and Jurečková (2000)).

As estimators of $s = \frac{1}{f(0)}$, we can use either (4.74) or (4.75).

6.3 Let $X_1,\dots,X_n$ be a sample from a population with distribution function $F(x - \theta)$, and F symmetric around 0. The α-trimmed mean $T_n(\alpha)$, defined in Example 3.2 is asymptotically normally distributed,

$$\mathcal{L}\left\{\sqrt{n}(T_n(\alpha) - \theta)\right\} \to \mathcal{N}(0, \sigma^2(\alpha, F))$$

where

$$\sigma^2(\alpha, F) = \frac{1}{(1-2\alpha)^2}\left\{\int_{F^{-1}(\alpha)}^{F^{-1}(1-\alpha)} x^2 dF(x) + 2\alpha(F^{-1}(1-\alpha))^2\right\}$$

Let α_F be a value minimizing $\sigma^2(\alpha, F)$. It is unknown, provided that F is unknown. The estimation of α_F (adaptive choice of the trimmed mean) was considered by Tukey and McLaughlin (1963), Jaeckel (1971), and Hall (1981).

6.4 The α-trimmed mean is asymptotically efficient for f in the Bahadur sense if and only if the central part of distribution F, namely $100(1 - 2\alpha)\%$, is normal, while the distribution has $100\alpha\%$ of exponential tails on both sides (Fu (1980)). This can be explained so that the problem of estimating α is equivalent to estimating the proportion of nonnormality from the observations.

6.5 Generate samples of a linear regression model with different distribution of errors and apply the function LADLS to these data.

6.6 Apply the function LADLS on the dataset used in Chapter 4 (the Hertzsprung-Russell diagram of the Star Cluster starsCYG, father.son dataset).

6.7 Write an R procedure that computes the residuals for $LSLAD$-estimator.

6.8 Apply the functions LADLS.test from Appendix A to testing of the regression coefficient on the datasets of Chapter 4.

Chapter 7

Robust and nonparametric procedures in measurement error models

7.1 Introduction

Measurement technologies are often affected by random errors, which make the estimates of parameters biased, and thus inconsistent. This problem appears in analytic chemistry, in environmental monitoring, in modeling astronomical data, in biometrics, and practically in all parts of reality. Moreover, some observations can be undetected, e.g., when the measured flux (light, magnetic) in the experiment falls below some flux limit. In econometrics, the errors can arise due to misreporting, miscoding by the collectors of the data, or by incorrect transformation from initial reports. Analytic chemists try to construct a calibration curve and to interpolate the correct unknown sample. However, even the calibration can be affected by measurement errors. The mismeasurements make the statistical inference biased, and they distort the trends in the data.

Measurement errors often appear also in the regression models, where either the regressor or the response or both can be affected by random errors. The mismeasurements make the statistical inference biased, and they distort the trends in the data. Technicians, geologists and other specialists are aware of this problem. They try to reduce the bias of their estimators with various ad hoc procedures; however, the bias cannot be substantially reduced or even completely eliminated, unless we have some additional knowledge on the behavior of measurement errors.

Practical aspects of measurement error models are described Akritas and Bershady (1996), Hyk and Stojek (2013), Kelly (2007), Marques (2004), Rocke and Lorenzato (1995), Zhang et al. (2018), among others. Among the monographs on the statistical inference in the error-in-variables (EV) models, we should mention those by Fuller (1987), by Carroll et al. (2006), and by Cheng and van Ness (1999), among others. While the monographs by Fuller (1987) and by Cheng and van Ness (1999) deal mostly with the classical Gaussian set up, Carroll et al. (2006) discusses numerous inference procedures under a semi-parametric setup. Nonparametric methods in EV models were studied by Carroll et al. (1999), Carroll et al. (2007) (see also their references), and further by Fan and Truong (1993), among others. The regression quantile theory in the area of EV models was started by He and Liang (2000). Adcock (1877) was the first who realized the importance of the situation. Arias et al. (2001) used an instrumental variable estimator for quantile regression, and Hausman (2001) and Hyslop and Imbens (2001) described the recent developments in treating the effect of mismeasurement on econometric models.

Regarding the existing rich literature on measurement errors, we shall concentrate on the nonparametric methods, and among them on methods based on ranks. The advantage of *rank and signed rank procedures* in the measurement error models was discovered recently in Jurečková et al. (2001), Navrátil and Saleh (2011), Navrátil (2012), Saleh et al. (2012) and in Sen et al. (2013). There it is shown that the critical region of the rank test in the linear model is insensitive to measurement errors under very general conditions; the measurement errors affect only the power of the test, not its size. However, the R-estimators can be also biased, because their distributions originate from the power of the test. Nevertheless, the R-estimates are still less sensitive to measurement errors than other estimates: As it is shown in Jurečková et al. (2016), the local asymptotic bias of the R-estimator of the regression parameter does not depend on the unknown distribution of the model errors. If only the covariates are affected by measurement errors, one can try to reduce the bias by a suitable robust calibration procedure, e.g., one constructed in the thesis of I. Müller (1996); it works in the multivariate regression model, when the dimension of the response is comparable with the dimension of the covariate.

7.2 Types of measurement errors, misspecification and violation of assumptions

Let us illustrate the situation on a *semiparametric partially linear* model

$$Y_i = \beta_0 + \mathbf{x}_i^\top \boldsymbol{\beta} + h(\mathbf{Z}_i) + U_i, \quad i = 1, \ldots, n, \tag{7.1}$$

where a real response Y is regressed to observable covariates $\mathbf{x}$ and further it depends on some possibly unobservable $\mathbf{Z}$. Here the $\mathbf{x}_i$ are known (nonstochastic) p-vectors, not all the same, $\mathbf{Z}_i$ is a stochastic q-vector covariate ($q \geq 1$), and the form of the function $h(\mathbf{Z}_i)$ is unspecified. The $\mathbf{Z}_i$ may be observable, partially observable or unobservable.

The statistical interest is often confined to the parameter $\boldsymbol{\beta}$, regarding $h(\cdot)$ as a nuisance function. Although $h(\cdot)$ is unspecified, it is of interest to distinguish whether the covariate $\mathbf{Z}$ is observable or not. If it is unobservable, (7.1) corresponds to the *latent effects* model. If the $\mathbf{Z}_i$ are observable (but eventually with measurement error) and are regarded as identically distributed random variables with some unspecified distribution (and independent of the error U_i), then $U_i^* = U_i + h(\mathbf{Z}_i)$, $i = 1, \ldots, n$ are i.i.d. random variables with unknown distribution function.

As an illustration, let us mention Nummi and Möttönen (2004), who describe the computer-based forest harvesting technique in Scandinavia, where the tree stems are converted into smaller logs and the stem height and diameter measurements are taken at fixed intervals. The harvester receives the length and diameter data at the ith stem point from a sensor, and measuring and computing equipment enable a computer-based optimization of crosscutting. Nummi and Möttönen (2004) consider the regression dependence of the stem diameter measurement y_i on the stem height measurement x_i at the ith stem point, $i = 1, \ldots, n$. The problem of interest is the prediction for y_i and the testing of hypotheses on the parameters of the model; but both the stem diameter and the stem height contain measurement errors.

Such problems are described by the partially linear regression model (7.1): $\mathbf{x}_i$ is a p-vector covariate, $\mathbf{Z}_i$ is a q-vector covariate, the function $h(\cdot)$ is unknown, and the model error U_i is independent of $(\mathbf{x}_i, \mathbf{Z}_i)$, $i = 1, \ldots, n$. It means that the response variable Y_i depends on variable $\mathbf{x}_i$ in a linear way but is still related to other independent variables $\mathbf{Z}_i$ in an unspecified form, $i = 1, \ldots, n$. This model, admitting the measurement errors, is flexible and enables us to model various situations, including latent variables.

Many practical problems lead to the model (7.1), and the interest is often in the inference on parameter $\boldsymbol{\beta}$, either testing or an estimation. For testing we strongly recommend nonparametric tests based on rank statistics. We often want to test the null hypothesis of no (partial) regression of Y on $\mathbf{x}$, treating β_0 and $h(\cdot)$ as nuisance parameters. Then we have the hypothesis

$$\mathbf{H}_0 : \ \boldsymbol{\beta} = \mathbf{0} \quad vs \quad \mathbf{H}_1 : \ \boldsymbol{\beta} \neq \mathbf{0} \tag{7.2}$$

with nuisance β_0 and $h(\cdot)$.

Although $h(\cdot)$ is unspecified in (7.1), the covariate $\mathbf{Z}$ can still be observable; then (7.1) corresponds to the *latent effects* model. Moreover, if the $\mathbf{Z}_i$ are observable and regarded as identically distributed random variables, independent of the error e_i, then we can put $e_i^* = e_i + h(\mathbf{Z}_i)$, which are other i.i.d. random variables with unknown distribution function. Then we are naturally led to nonparametric tests based on rank statistics. We can test not only the hypothesis (7.2) in model (7.1), but we can admit a more general model

$$\begin{aligned} Y_i &= \beta_0 + \mathbf{x}_i^\top \boldsymbol{\beta} + \mathbf{x}_i^{*\top} \boldsymbol{\beta}^* + h(\mathbf{Z}_i) + U_i \\ &= \beta_0 + \mathbf{x}_i^\top \boldsymbol{\beta} + \mathbf{x}_i^{*\top} \boldsymbol{\beta}^* + U_i^*, \ i = 1, \ldots, n \end{aligned} \tag{7.3}$$

and test for the hypothesis $\mathbf{H}_{01} : \ \boldsymbol{\beta} = \mathbf{0}$ with β_0, $\boldsymbol{\beta}^*$, $h(\cdot)$ nuisance, even if the $\mathbf{x}_i$ are affected by additive errors, but not the $\mathbf{x}_i^*$. The tests are based on regression rank scores and are asymptotically distribution-free. (Jurečková et al. (2010)), even if the $\mathbf{Z}_i$ are observable but subject to measurement errors (as in Nummi and Möttönen (2004)). This is due to the fact that the e_i^* are still i.i.d. random variables, though possibly with another distribution function.

Among the estimators in the measurement error situation, we can also recommend the R-estimators; they are still less sensitive to measurement errors than other estimates.

7.3   Measurement errors in nonparametric testing

Consider the linear regression model

$$Y_i = \beta_0 + \mathbf{x}_{ni}^\top \boldsymbol{\beta} + U_i, \quad i = 1, \ldots, n \tag{7.4}$$

with observations $Y_1, \ldots, Y_n$, independent errors $U_1, \ldots, U_n$, identically or non-identically distributed with unknown distribution functions. The measurement errors can appear in the responses Y_i, in the covariates $\mathbf{x}_i$, or in both. Either case can have a different effect on the test of a hypothesis about parameter $\boldsymbol{\beta}$. We shall consider these situations separately, and concentrate on the rank tests, which show the smallest sensitivity.

7.3.1   Rank tests with measurement errors in responses

Consider the situation that the responses $Y_1, \ldots, Y_n$ are affected by random errors, and instead of Y_i we only observe $W_i = Y_i + V_i$, $i = 1, \ldots, n$. We want to test the hypothesis $\mathbf{H} : \boldsymbol{\beta} = \mathbf{0}$ against the alternative $\mathbf{K} : \boldsymbol{\beta} \neq \mathbf{0}$, with β_0 unknown. Our interest is to find how the eventual measurement errors in Y_i's affect the behavior and the efficiency of the rank tests of the linear hypothesis in the linear regression model. In other words, we are interested in what happens if we ignore the presence of eventual measurement errors and use the ordinary rank tests.

Let us assume that $U_{n1}, \ldots, U_{nn}$ are identically distributed. Then the rank tests of $\mathbf{H} : \boldsymbol{\beta} = \mathbf{0}$ perform well and keep the size α even under measurement errors; a change can appear only in the power of the test. Denote $\bar{\mathbf{X}}_n$ as the matrix with the rows $\mathbf{x}_{ni} - \bar{\mathbf{x}}_n$, $i = 1, \ldots, n$. Assume that it satisfies

$$\mathbf{Q}_n = \frac{1}{n}\bar{\mathbf{X}}_n^\top \bar{\mathbf{X}}_n = \frac{1}{n}\sum_{i=1}^n (\mathbf{x}_{ni} - \bar{\mathbf{x}}_n)(\mathbf{x}_{ni} - \bar{\mathbf{x}}_n)^\top \to \mathbf{Q} \text{ as } n \to \infty \tag{7.5}$$

$$n^{-1} \max_{1 \leq i \leq n} \left\{ (\mathbf{x}_{ni} - \bar{\mathbf{x}}_n)^\top \mathbf{Q}_n^{-1}(\mathbf{x}_{ni} - \bar{\mathbf{x}}_n) \right\} \to 0 \text{ as } n \to \infty$$

where $\mathbf{Q}$ is a positively definite $p \times p$ matrix. We assume that the distribution function F of the errors U_i has an absolutely continuous density f and finite Fisher information $\mathcal{I}(f) = \int_{\mathbb{R}} \left(\frac{f'(z)}{f(z)} \right)^2 dF(z) < \infty$.

Instead of Y_{ni} we can only observe $W_{ni} = Y_{ni} + V_{ni}$, $i = 1, \ldots, n$, where $V_{n1}, \ldots, V_{nn}$ are i.i.d. random measurement errors, independent of the Y_{ni} and x_{ni}. Their joint distribution (say G) is unknown, and we only assume that it has an absolutely continuous density g.

The ranks $R_{n1}, \ldots, R_{nn}$ of $Y_{n1}, \ldots, Y_{nn}$ are not directly observable, and we only observe the ranks $Q_{n1}, \ldots, Q_{nn}$ of $W_{n1}, \ldots, W_{nn}$. Notice that $W_{n1}, \ldots, W_{nn}$ are independent and identically distributed under the hypothesis $\mathbf{H}$; hence

$$P\Big((Q_{n1}, \ldots, Q_{nn}) = (r_1, \ldots, r_n)\Big) = \frac{1}{n!}$$

for every permutation $(r_1, \ldots, r_n)$ of $1, \ldots, n$. Denote H and h as the joint distribution function and density of $W_{n1}, \ldots, W_{nn}$. Their Fisher information satisfies $\mathcal{I}(h) \leq \mathcal{I}(f)$ and hence it is finite (see Hájek et al. (2000), Theorem I.2.3).

Observing the ranks $Q_{n1}, \ldots, Q_{nn}$, we base our test on the vector of linear rank statistic

$$\widetilde{\mathbf{S}}_n = \sum_{i=1}^n (\mathbf{x}_{ni} - \bar{\mathbf{x}}_n) a_n(Q_{ni}) = \sum_{i=1}^n (\mathbf{x}_{ni} - \bar{\mathbf{x}}_n) \varphi \left(\frac{Q_{ni}}{n+1} \right) \qquad (7.6)$$

where φ is a nondecreasing score function, square-integrable on $(0,1)$, for simplicity such that $\int_0^1 \varphi(u) du = 0$. The test criterion for $\mathbf{H}$ is the quadratic form in $\mathbf{S}_n$,

$$\mathcal{T}_n^2 = (A(\varphi))^{-2} \, \mathbf{S}_n^\top \mathbf{Q}_n^{-1} \mathbf{S}_n, \qquad (7.7)$$

$$A^2(\varphi)) = \int_0^1 \varphi^2(u) du$$

and its asymptotic distribution under $\mathbf{H}$ is χ^2 with p degrees of freedom. Hence, we reject hypothesis $\mathbf{H}$ in favor of $\mathbf{K}$ if $\mathcal{T}_n^2$ exceeds the $100(1-\alpha)$ quantile of the χ^2 with p degrees of freedom.

Because the distribution of $\widetilde{\mathbf{S}}_n$, and thus also that of $\mathcal{T}_n^2$ are distribution-free under $\mathbf{H}$, the test based on $\mathcal{T}_n^2$ has the same probability of the error of the 1st kind as if there are no measurement errors. However, the measurement errors affect the efficiency/power of the test. We shall illustrate the loss of efficiency on the regression line $Y_i = \beta_0 + x_{in}\beta + U_i$, $i = 1, \ldots, n$ with regressors $(x_{n1}, \ldots, x_{nn})$ satisfying the Noether condition

$$\lim_{n \to \infty} \left[\max_{1 \leq i \leq n} (x_{ni} - \bar{x}_n)^2 \left(\sum_{j=1}^n (x_{nj} - \bar{x}_n)^2 \right)^{-1} \right] = 0, \quad \bar{x}_n = \frac{1}{n} \sum_{i=1}^n x_{ni}. \quad (7.8)$$

Moreover, assume that

$$\lim_{n \to \infty} C_n^2 = C^2 > 0, \quad \text{where} \quad C_n^2 = \frac{1}{n} \sum_{j=1}^n (x_{nj} - \bar{x}_n)^2, \quad (7.9)$$

hence $\max_{1 \leq i \leq n} |x_{ni} - \bar{x}_n| = o(n^{1/2})$.

The test of $\mathbf{H} : \beta = 0$ is based on the linear rank statistic

$$\widetilde{S}_n = \sum_{i=1}^n (x_{ni} - \bar{x}_n) a_n(Q_{ni}) = \sum_{i=1}^n (x_{ni} - \bar{x}_n) \varphi \left(\frac{Q_{ni}}{n+1} \right) \qquad (7.10)$$

where Q_{ni} is the rank of $W_i = Y_i + V_i$, $i = 1, \ldots, n$. The test rejects $\mathbf{H}$ in favor of $\beta > 0$ if $\widetilde{S}_n > K_\alpha$ where K_α is the critical value such that

$$P_H \left(\widetilde{S}_n > K_\alpha \right) + \gamma P_H \left(\widetilde{S}_n = K_\alpha \right) = \alpha, \quad 0 \leq \gamma < 1.$$

By Hájek et al. (2000), under validity of $\mathbf{H}$ we have the following approximation for $\tilde{S}_n$:

$$\lim_{n \to \infty} \mathrm{E}_{H_0} \left(\sigma_n^{-2} (\tilde{S}_n - \tilde{T}_n)^2 \right) = 0, \qquad (7.11)$$

$$\sigma_n^2 = \frac{1}{n} \sum_{i=1}^n (x_{ni} - \bar{x}_n)^2 A^2(\varphi)$$

where

$$\tilde{T}_n = n^{-1/2} \sum_{i=1}^n (x_{ni} - \bar{x}_n) \varphi(H(Y_{ni})) = n^{-1/2} \sum_{i=1}^n (x_{ni} - \bar{x}_n) \varphi(J_{ni})$$

and $J_{n1}, \ldots, J_{nn}$ is the random sample from the uniform $(0,1)$ distribution. Under the local alternative $\mathbf{K}_n$ to $\mathbf{H}$,

$$\mathbf{K}_n : \ \beta = \beta_n = n^{-1/2} \beta^*, \ \beta^* > 0, \ \text{ we still have}$$

$$\sigma_n^{-1} (\tilde{S}_n - \tilde{T}_n) \xrightarrow{P} 0 \quad \text{as } n \to \infty. \qquad (7.12)$$

Hence, the asymptotic distributions of $\tilde{S}_n$ under $\mathbf{K}_n$ coincide with those of $\tilde{T}_n$, under $\mathbf{H}$ as well as under $\mathbf{K}_n$, when they are normal

$$\mathcal{N} \left(C^2 \, \gamma(\varphi, f) \, \beta^*, \ C^2 A^2(\varphi) \right) \ \text{ and } \ \mathcal{N} \left(C^2 \, \gamma(\varphi, h) \, \beta^*, \ C^2 A^2(\varphi) \right),$$

respectively, where

$$\gamma(\varphi, h) = \int_0^1 \varphi(t) \varphi(t, h) dt, \quad \varphi(t, h) = -\frac{h'(H^{-1}(t))}{h(H^{-1}(t))}, \quad 0 < t < 1$$

and similar for $\gamma(\varphi, f)$. Hence, the asymptotic relative efficiency of the test based on $\tilde{S}_n$ with respect to that based on S_n is

$$e(\tilde{S}_n, S_n) = \left(\frac{\gamma(\varphi, h)}{\gamma(\varphi, f)} \right)^2 = \left(\frac{\int_0^1 h(H^{-1}(t)) d\varphi(t)}{\int_0^1 f(F^{-1}(t)) d\varphi(t)} \right)^2. \qquad (7.13)$$

In the special case of the Wilcoxon test, $e(\tilde{S}_n, S_n) \leq 1$, and the equality appears if and only if there are measurement errors with probability 0. Actually,

$$\int_0^1 h(H^{-1}(t)) dt = \int_{\mathbb{R}} h^2(z) dz = \int_{\mathbb{R}} \left(\int_{\mathbb{R}} f(z - u) dG(u) \right)^2 dz$$

$$\leq \int_{\mathbb{R}} \int_{\mathbb{R}} f^2(z - u) dG(u) dz = \int_{\mathbb{R}} f^2(z) dz = \int_0^1 f(F^{-1}(t)) dt.$$

Particularly, if the test is asymptotically optimal for f, i.e., if $\varphi(t) = \varphi(t, f)$, $0 < t < 1$, then $e(\tilde{S}_n, S_n) \leq \frac{\mathcal{I}(h)}{\mathcal{I}(f)} \leq 1$. In the general case,

$$e(\tilde{S}_n, S_n) \leq \frac{\mathcal{I}(h) A^2(\varphi)}{\gamma^2(\varphi, f)}.$$

On the other hand, we can use the test (7.10) even if the model errors are not equally distributed, but their distribution functions differ only in small distortions. As an example we can mention the model of regression line

$$Y_{ni} = \beta_0 + x_{ni}\beta + \sigma_{ni}U_{ni}, \ \sigma_{ni} = \exp\{n^{-1/2}z_{ni}\delta\}, \ i = 1,\ldots,n$$

with locally heteroscedastic errors, with given $z_{ni} \in \mathbb{R}_1$ and with an unknown parameter δ. Then the limiting probability of the critical region

$$\left\{(CA(\varphi))^{-1}\tilde{S}_n > \Phi^{-1}(1-\alpha)\right\}$$

under $\mathbf{H}$ is still equal to α, provided the density f of U_1 is symmetric around 0 and the score function φ is symmetric around $\frac{1}{2}$ (see Jurečková and Navrátil (2013)).

7.3.2   Rank tests with measurement errors in covariates

In the model (7.4), consider the situation that the measurement of the co-variates can be affected by random errors. Then, instead of x_{ij}, we observe $w_{ij} = x_{ij} + v_{ij}, \ i = 1,\ldots,n, \ j = 1,\ldots,p$, where $\mathbf{v}_{ni} = (v_{i1},\ldots,v_{ip})^\top, \ i = 1,\ldots,n$ are random measurement errors, which are not under our control. We assume that random vectors $\mathbf{v}_{ni}, \ i = 1,\ldots,n$ are independent and identically distributed and independent of $U_1,\ldots,U_n$, but their distribution is unknown. We are interested in the rank tests of the hypothesis $\mathbf{H}: \ \boldsymbol{\beta} = \mathbf{0}$ [or $\boldsymbol{\beta} = \boldsymbol{\beta}_0$]. It turns out that the rank tests keep the size α even with finite n, while measurement errors only affect their powers.

Denote $\bar{\mathbf{X}}_n$ as the matrix with the rows $\mathbf{x}_{ni} - \bar{\mathbf{x}}_n, \ i = 1,\ldots,n$, and let it satisfy

$$\mathbf{Q}_n = \frac{1}{n}\bar{\mathbf{X}}_n^\top\bar{\mathbf{X}}_n = \frac{1}{n}\sum_{i=1}^n (\mathbf{x}_{ni} - \bar{\mathbf{x}}_n)(\mathbf{x}_{ni} - \bar{\mathbf{x}}_n)^\top \to \mathbf{Q} \ \text{ as } \ n \to \infty,$$

$$n^{-1}\max_{1\leq i\leq n}\left\{(\mathbf{x}_{ni} - \bar{\mathbf{x}}_n)^\top\mathbf{Q}_n^{-1}(\mathbf{x}_{ni} - \bar{\mathbf{x}}_n)\right\} \to 0 \ \text{ as } \ n \to \infty \qquad (7.14)$$

where $\mathbf{Q}$ is a positively definite $p \times p$ matrix. We assume that the distribution function F of the errors has an absolutely continuous density f and finite Fisher information.

In the model without measurement errors, the rank test of $\mathbf{H}: \ \boldsymbol{\beta} = \mathbf{0}$ is based on the vector of linear rank statistics $\mathbf{S}_n \in \mathbb{R}^p$,

$$\mathbf{S}_n = n^{-1/2}\sum_{i=1}^n (\mathbf{x}_{ni} - \bar{\mathbf{x}}_n)a_n(R_{ni})$$

where $R_{n1},\ldots,R_{nn}$ are the ranks of $Y_1,\ldots,Y_n$ and $a_n(i)$ are again the scores generated by φ. The asymptotic distribution of $\mathbf{S}_n$ under $\mathbf{H}_0$ is normal $\mathcal{N}_p(\mathbf{0}, \ A^2(\varphi)\,\mathbf{Q})$. The test criterion for $\mathbf{H}$ is the quadratic form in $\mathbf{S}_n$

$$\mathcal{T}_n^2 = (A(\varphi))^{-2}\,\mathbf{S}_n^\top\mathbf{Q}_n^{-1}\mathbf{S}_n,$$

and its asymptotic null distribution is χ^2 with p degrees of freedom. The asymptotic distribution of $\mathcal{T}_n^2$ under the local alternative

$$\mathbf{K}_n : \ \beta = \beta_n = n^{-1/2}\beta^* \in \mathbb{R}^p, \tag{7.15}$$

that the model is $Y_i = \beta_0 + \mathbf{x}_{ni}^\top \beta_n + U_i, \ i = 1,\ldots,n$, is the noncentral χ^2 with p degrees of freedom and with the noncentrality parameter

$$\eta^2 = \beta^{*\top}\mathbf{Q}\beta^* \ \frac{\gamma^2(\varphi, F)}{A^2(\varphi)}. \tag{7.16}$$

However, while we observe Y_i's and their ranks, instead of $\mathbf{x}_{ni}$ we can observe only $\mathbf{w}_{ni} = \mathbf{x}_{ni} + \mathbf{v}_{ni}, \ i = 1,\ldots,n$, where $\mathbf{v}_{n1},\ldots,\mathbf{v}_{nn}$ are i.i.d. p-dimensional errors, independent of the U_i. Their distribution is generally unknown, but we should assume that they satisfy

$$\mathbf{V}_n = \frac{1}{n}\sum_{i=1}^{n}(\mathbf{v}_{ni} - \bar{\mathbf{v}}_n)(\mathbf{v}_{ni} - \bar{\mathbf{v}}_n)^\top \xrightarrow{P} \mathbf{V} \tag{7.17}$$

as $n \to \infty$, where $\mathbf{V}$ is a positively definite $p \times p$ matrix, for us unknown. Moreover, we assume that Y_i and $\mathbf{v}_i$ are independent, $i = 1,\ldots,n$, and that $(\mathbf{x}_{n1},\ldots,\mathbf{x}_{nn})$ and $(\mathbf{v}_{n1},\ldots,\mathbf{v}_{nn})$ are asymptotically uncorrelated, i.e.,

$$\frac{1}{n}\sum_{i=1}^{n}(\mathbf{v}_{ni} - \bar{\mathbf{v}}_n)^\top(\mathbf{x}_{ni} - \bar{\mathbf{x}}_{ni}) \xrightarrow{P} 0 \quad \text{as } n \to \infty. \tag{7.18}$$

Then (7.14), (7.17) and (7.18) imply that $\mathbf{Q}_n + \mathbf{V}_n$ are observable even under measurement errors, and as $n \to \infty$,

$$\mathbf{Q}_n + \mathbf{V}_n = \frac{1}{n}\sum_{i=1}^{n}(\mathbf{w}_{ni} - \bar{\mathbf{w}}_n)(\mathbf{w}_{ni} - \bar{\mathbf{w}}_n)^\top \xrightarrow{P} \mathbf{Q} + \mathbf{V} \tag{7.19}$$

If there are measurement errors, the test of $\mathbf{H} : \ \beta = \mathbf{0}$ can only be based on the observable entities; hence we use the test based on the vector of linear rank statistics $\tilde{\mathbf{S}}_n \in \mathbb{R}^p$,

$$\tilde{\mathbf{S}}_n = n^{-1/2}\sum_{i=1}^{n}(\mathbf{w}_{ni} - \bar{\mathbf{w}}_n)a_n(R_{ni}). \tag{7.20}$$

Under the above conditions, its asymptotic null distribution is normal $\mathcal{N}_p\left(\mathbf{0},\ A^2(\varphi)\,(\mathbf{Q} + \mathbf{V})\right)$, while under $\mathbf{K}_n$ of (7.15) it has the asymptotic p-variate normal distribution $\mathcal{N}_p\left(\gamma(\varphi, F)\,\mathbf{Q}\beta^*,\ A^2(\varphi)\,(\mathbf{Q} + \mathbf{V})\right)$.

The test criterion for $\mathbf{H}$ now is

$$\tilde{\mathcal{T}}_n^2 = (A(\varphi))^{-2}\ \tilde{\mathbf{S}}_n^\top(\mathbf{Q}_n + \mathbf{V}_n)^{-1}\tilde{\mathbf{S}}_n \tag{7.21}$$

whenever the matrix $(\mathbf{Q}_n + \mathbf{V}_n)$ is regular. It is calculable from the observed data, and its asymptotic distribution under the hypothesis $\mathbf{H}$ is χ^2 with p degrees of freedom. Under the local alternative $\mathbf{K}_n$, the asymptotic distribution of $\widetilde{\mathcal{T}}_n^2$ will be noncentral χ^2 with p degrees of freedom and with the noncentrality parameter

$$\tilde{\eta}^2 = \boldsymbol{\beta}^{*\top} \, \mathbf{Q}(\mathbf{Q} + \mathbf{V})^{-1}\mathbf{Q} \, \boldsymbol{\beta}^* \frac{\gamma^2(\varphi, F)}{A^2(\varphi)}. \tag{7.22}$$

The asymptotic relative efficiency of the test of $\mathbf{H}$ based on $\widetilde{\mathcal{T}}_n^2$ with respect to the test based on $\mathcal{T}_n^2$ then equals

$$ARE(\widetilde{\mathcal{T}} : \mathcal{T}) = \frac{\boldsymbol{\beta}^{*\top} \, \mathbf{Q}(\mathbf{Q} + \mathbf{V})^{-1}\mathbf{Q} \, \boldsymbol{\beta}^*}{\boldsymbol{\beta}^{*\top}\mathbf{Q}\boldsymbol{\beta}^*}. \tag{7.23}$$

Thus

$$\underline{\lambda} \leq ARE(\widetilde{\mathcal{T}} : \mathcal{T}) \leq \overline{\lambda}$$

where $\underline{\lambda}$, $\overline{\lambda}$ respectively stand for the minimum and maximum characteristic roots of $\mathbf{Q}(\mathbf{Q} + \mathbf{V})^{-1}$.

The main positive message is that we can construct a distribution-free rank test of hypothesis $\mathbf{H}_0 : \boldsymbol{\beta} = \mathbf{0}$ in the model even when the regression matrix $\mathbf{X}$ can only be measured with random additive errors. Its loss of efficiency is given in (7.23). We can use this test even if the responses Y_i are further damaged by additional measurement errors. As we have seen before, the significance level would not change, only we should count with an additional loss of efficiency.

7.3.3   Numerical illustration

Consider the linear regression model

$$Y_i = \beta_0 + x_i\beta_1 + e_i, \quad i = 1, \ldots, n$$

and the problem of testing the hypothesis $\mathbf{H} : \beta_1 = 0$ against alternative $\mathbf{K} : \beta_1 \neq 0$ in the situation where both the regressors and responses in the model are affected by random errors. Instead of Y_i we observe $Y_i + U_i$, $i = 1, \ldots, n$ and instead of $\mathbf{x}_i$ we observe $\mathbf{x}_i + \mathbf{v}_i$, $i = 1, \ldots, n$.

The test criterion 7.21 for $\mathbf{H}$ we could implement as follows:

```
> rank.test.meas<-function(x, y, score = "wilcoxon")
+ {
+ n<-length(x[,1])
+ xp<-apply(x,2,mean)
+ x2<-x-matrix(rep(xp,n),nrow=n,byrow=T)
+ Q=t(x2)%*%x2/n
```

```
+ por<-rank(y)
+ if(score == "wilcoxon") {
+ skor<-por/(n+1)-0.5
+ A2<-12
+ }
+   else if(score == "normal") {
+     skor<-qnorm(por/(n+1))
+     A2<-1
+ }
+ else if(score == "sign") {
+     skor<-0.5*sign(por/(n+1)-0.5)
+     A2<-1
+ }
+ sn<-n^(-0.5)*apply(x2*skor,2,sum)
+ t(sn)%*%solve(Q)%*%sn*A2
+ }
```

We shall illustrate the power of the Wilcoxon test, i.e., $\varphi(u) = 2u - 1$, $0 \leq u \leq 1$ and $A^2(\varphi) = \frac{1}{12}$, by means of the frequency of rejections under various error distributions and various values of β_1. The chosen value of the intercept is $\beta_0 = 1$. The regressors $x_1, \ldots, x_n$ were simulated for $n = 20$ and 500 from the uniform distribution $U[-2, 10]$, independently of the errors. The measurement errors U_i and v_i, $i = 1, \ldots, n$ were generated independently from the normal and uniform distributions with various parameters.

A total of 100 000 replications were simulated and the test on level $\alpha = 0.05$ was performed every time; the mean power was then calculated. The results of the simulation are presented in Figures 7.1–7.2; they show a good performance of the test for the measurement errors with small variance.

With the following code we can simulate and display the power of the mentioned tests.

```
x20<-runif(20,-2,10)
res<-matrix(0,10000,5)
res2<-matrix(0,11,5)
for (kk in -5:5){
bet<-kk/10
for (i in 1:10000) {
 e<-rnorm(20)
 yy<-rep(1,20)+x20*bet+e
 u20a<-rnorm(20,0,0.7)
 u20b<-rnorm(20,0,1)
 u20c<-rnorm(20,0,2)
 u20d<-runif(20,-1,1)
 v20a<-rnorm(20,0,0.7)
 v20b<-rnorm(20,0,1)
```

```
v20c<-rnorm(20,0,2)
v20d<-runif(20,-1,1)
res[i,1]<-rank.test.meas(as.matrix(x20),yy)
res[i,2]<-rank.test.meas(as.matrix(x20+u20a),yy+v20a)
res[i,3]<-rank.test.meas(as.matrix(x20+u20b),yy+v20b)
res[i,4]<-rank.test.meas(as.matrix(x20+u20c),yy+v20c)
res[i,5]<-rank.test.meas(as.matrix(x20+u20d),yy+v20d)
}
for(i in 1:5) res2[kk+6,i]<-1-length(res[(res[,i]<qchisq
          (0.975,1))& (res[,i]>qchisq(0.025,1)),i])/10000
}
#
plot((-5:5)/10,res2[,1],type="l",ylim=c(0,1),xlab="beta_1",
          ylab="simulated power",lty=1)
abline(h=0.05,lty=1)
lines((-5:5)/10,res2[,2],lty=2)
lines((-5:5)/10,res2[,3],lty=3)
lines((-5:5)/10,res2[,5],lty=4)
```

7.4   Measurement errors in nonparametric estimation

We shall now devote some time to the estimation problem in model (7.4) if there are measurement errors. The methods based on ranks are flexible in testing hypotheses; so we are interested in the R-estimator of the slope vector $\boldsymbol{\beta}$, considering β_0 as nuisance parameter. The regressors $\mathbf{x}_{ni}$ are either deterministic or random, but affected by additive random measurement errors, so that instead of $\mathbf{x}_{ni}$ we measure $\mathbf{w}_{ni} = \mathbf{x}_{ni} + \mathbf{v}_{ni}$, $i = 1, \ldots, n$, where $\mathbf{v}_{n1}, \ldots, \mathbf{v}_{nn}$ are p-dimensional random errors, identically distributed with an unknown distribution function. Moreover, there are additive measurement errors in the responses, thus instead of Y_{ni} we observe $Y_{ni}^* = Y_{ni} + z_{ni}$, where $z_{n1}, \ldots, z_{nn}$ are i.i.d. random variables.

If we are not aware of measurement errors, we in fact work with an R-estimator $\widehat{\boldsymbol{\beta}}_n$ of $\boldsymbol{\beta}$ in the contaminated model

$$Y_{ni}^* = \beta_0 + \mathbf{w}_{ni}^\top \boldsymbol{\beta} + U_{ni}^*, \ i = 1, \ldots, n, \tag{7.24}$$

where $U_{ni}^* = U_{ni}^*(\boldsymbol{\beta}) = U_{ni} + z_{ni} - \mathbf{v}_{ni}^\top \boldsymbol{\beta}$, $i = 1, \ldots, n$ are i.i.d random variables. Let $R_{ni}(\mathbf{b}) = R_{ni}(\mathbf{b}, \boldsymbol{\beta})$ be the rank of the residual

$$
\begin{aligned}
Y_{ni}^* - \mathbf{w}_{ni}^\top \mathbf{b} &= U_{ni} + z_{ni} + \mathbf{x}_{ni}^\top \boldsymbol{\beta} - \mathbf{w}_{ni}^\top \mathbf{b} \\
&= U_{ni} + u_{ni} - \mathbf{w}_{ni}^\top \mathbf{b}^* - \mathbf{v}_{ni}^\top \boldsymbol{\beta}, \quad i = 1, \ldots, n
\end{aligned}
$$

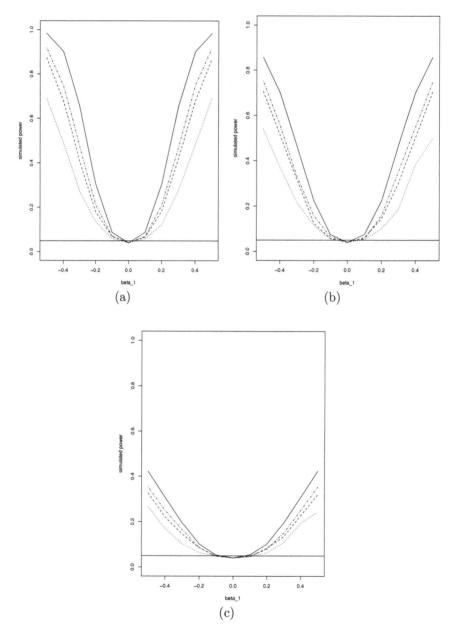

Figure 7.1 *Mean power of the Wilcoxon test for $n = 20$, for standard normal errors e_i, $i = 1, \ldots, n$ (a), Laplace distribution of errors (b) and Cauchy distribution (c). Solid line corresponds to the standard test without the measurement errors. The situations where $Y_1, \ldots, Y_n$ and $x_1, \ldots, x_n$ are affected by random errors are denoted by the dashed line (both normal distribution N[0,0.7]), the dotted line (both normal distribution N[0,1]) and the dot-dash line (both uniform U[-1,1]).*

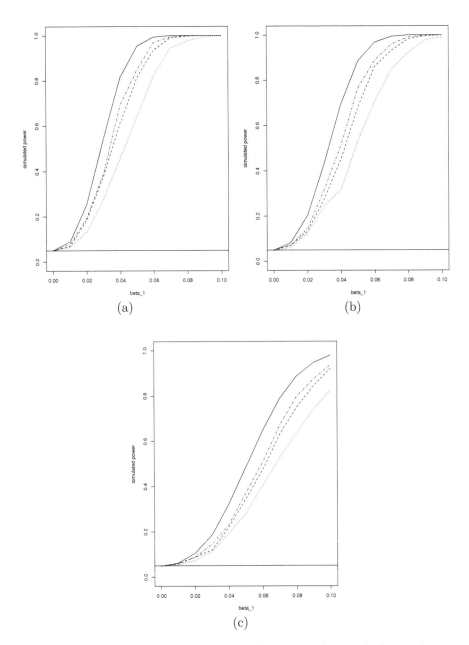

Figure 7.2 *Mean power of the Wilcoxon test for $n = 500$, for standard normal errors e_i, $i = 1, \ldots, n$ (a), Laplace distribution of errors (b) and Cauchy distribution (c). Solid line corresponds to the standard test without the measurement errors. The situations where $Y_1, \ldots, Y_n$ and $x_1, \ldots, x_n$ are affected by random errors are denoted by the dashed line (both normal distribution N[0,0.7]), the dotted line (both normal distribution N[0,1]) and the dot-dash line (both uniform U[-1,1]).*

where $\mathbf{b}^* = \mathbf{b} - \boldsymbol{\beta}$. We define the rank estimator $\widehat{\boldsymbol{\beta}}_n$ of $\boldsymbol{\beta}$ as the minimizer of the Jaeckel (1971) measure of the rank dispersion

$$
\begin{aligned}
\mathcal{D}_n(\mathbf{b}) &= \sum_{i=1}^{n} \left[Y_{ni}^* - \mathbf{w}_{ni}^\top \mathbf{b} \right] \left(a_n(R_{ni}(\mathbf{b})) - \bar{a}_n \right) \\
&= \sum_{i=1}^{n} \left[e_{ni} + u_{ni} - (\mathbf{w}_i - \bar{\mathbf{w}}_n)^\top \mathbf{b}^* - (\mathbf{v}_i - \bar{\mathbf{v}}_n)^\top) \boldsymbol{\beta} \right] \\
&\qquad \left(a_n(R_i(\mathbf{b}, \boldsymbol{\beta})) - \bar{a}_n \right) \tag{7.25}
\end{aligned}
$$

with respect to $\mathbf{b} \in \mathbb{R}^p$. Jaeckel showed that $\mathcal{D}_n(\mathbf{b})$ is convex and piecewise linear in $\mathbf{b} \in \mathbb{R}^p$. He also showed that the subgradient of $\mathcal{D}_n(\mathbf{b})$ is $-n^{1/2}\mathbf{S}_n(\mathbf{b})$, where

$$
\mathbf{S}_n(\mathbf{b}) = (S_{nj}(\mathbf{b}); \ j = 1, \dots, p)^\top = n^{-1/2} \sum_{i=1}^{n} (\mathbf{w}_{ni} - \bar{\mathbf{w}}_n) a_n(R_{ni}(\mathbf{b}))
$$

is the linear rank statistic; thus the minimizer of $\mathcal{D}$ exists. In the absence of measurement errors, i.e., if $\mathbf{w}_{ni} = \mathbf{x}_{ni}$, $z_{ni} = 0$, $i = 1, \cdots, n$, the estimator $\widehat{\boldsymbol{\beta}}_n$ is consistent and asymptotically normal. However, $\widehat{\boldsymbol{\beta}}_n$ is biased in the presence of measurement errors, and so is every estimator of $\boldsymbol{\beta}$, unless we know the distribution of the measurement errors. If only the covariates are affected by measurement errors, one can sometimes try to reduce the bias by a suitable robust calibration procedure: It estimates the value of the covariate from the value of the response and from a preliminary estimator of the regression parameter. Robust calibration procedures work in the multivariate regression model, when the dimension of the response is comparable with the dimension of the regressor (see e.g., I. Müller (1996)).

We recall the function `rfit` mentioned in Chapter 4 to minimize the Jaeckel's dispersion function 7.25.

```
library(quantreg)
library(Rfit)
rfit(formula)
```

The local asymptotic bias of $\widehat{\boldsymbol{\beta}}_n$ under $\boldsymbol{\beta}$ close to $\mathbf{0}$, i.e., under

$$
\boldsymbol{\beta} = \boldsymbol{\beta}_n = n^{-1/2}\boldsymbol{\beta}^* \quad \text{with a fixed } \boldsymbol{\beta}^* \in \mathbb{R}^p, \tag{7.26}
$$

was derived in Jurečková et al. (2016) and it is described in Theorem 7.1 below. The local asymptotic bias of the R-estimator is distribution-free, and it surprisingly depends neither on the chosen rank test (on its score-generating function) nor on the parent distribution F of the model errors (which is unknown). It is proven under some conditions on distributions of U_{ni}, v_{ni}, z_{ni} and on matrix $\mathbf{X}_n$. Usually, all measurement errors are supposed to be mutually independent and independent uf $U_{ni,i=1,\dots,n}$ and similarly for regressors, if they are random. Moreover,

$$\mathbb{E} \, \mathbf{V}_n \to \mathbf{V} \quad \text{where} \quad \mathbf{V}_n = n^{-1} \sum_{i=1}^{n} (\mathbf{v}_{ni} - \bar{\mathbf{v}}_n)(\mathbf{v}_{ni} - \bar{\mathbf{v}}_n)^\top$$

and $\mathbf{V}$ is a positive definite $p \times p$ matrix;

$$\mathbb{E} \left[n^{-1} \sum_{i=1}^{n} (\mathbf{v}_{ni} - \bar{\mathbf{v}}_n)(\mathbf{x}_{ni} - \bar{\mathbf{x}}_n)^\top \right] \to \mathbf{0}.$$

If the regressors $\mathbf{x}_{ni}$ are nonrandom, then

$$\mathbf{Q}_n = \frac{1}{n} \sum_{i=1}^{n} (\mathbf{x}_{ni} - \bar{\mathbf{x}}_n)(\mathbf{x}_{ni} - \bar{\mathbf{x}}_n)^\top \to \mathbf{Q},$$

and $\mathbf{Q}$ is positively definite $p \times p$ matrix. Moreover,

$$\frac{1}{n} \max_{1 \le i \le n} (\mathbf{x}_{ni} - \bar{\mathbf{x}}_n)^\top (\mathbf{Q}_n)^{-1} (\mathbf{x}_{ni} - \bar{\mathbf{x}}_n) \to 0.$$

If the regressors $\mathbf{x}_{ni}$ are random, then

$$\mathbb{E} \left[\frac{1}{n} \sum_{i=1}^{n} (\mathbf{x}_{ni} - \bar{\mathbf{x}}_n)(\mathbf{x}_{ni} - \bar{\mathbf{x}}_n)^\top \right] \to \mathbf{Q},$$

where $\mathbf{Q}$ is positively definite $p \times p$ matrix.

Finally let $m(\cdot)$, $M(\cdot)$ denote the density and distribution function of $U_{ni} + z_{ni}$, $i = 1, \cdots, n$. We denote

$$\gamma_m = -\int_{\mathbb{R}^1} \varphi(M(z))dm(z), \qquad A_m^2(\varphi) = \gamma_m^{-2} \int_0^1 \varphi^2(u)du, \quad (7.27)$$
$$\mathbf{B} = -(\mathbf{Q} + \mathbf{V})^{-1}\mathbf{V}\,\boldsymbol{\beta}^0.$$

Theorem 7.1 *Under the above conditions and under the local alternatives (7.26), the R-estimator $\widehat{\boldsymbol{\beta}}_n$ is asymptotically normally distributed with the bias $\mathbf{B} = -(\mathbf{Q} + \mathbf{V})^{-1}\mathbf{V}\,\boldsymbol{\beta}^0$, i.e.,*

$$n^{1/2}(\widehat{\boldsymbol{\beta}}_n - \boldsymbol{\beta}_n) \xrightarrow{\mathcal{D}} \mathcal{N}_p \left(\mathbf{B}, (\mathbf{Q} + \mathbf{V})^{-1}A^2(\varphi) \right). \qquad (7.28)$$

Theorem 7.1 is proven in Jurečková et al. (2016), along with the following Corollary:

Corollary 7.1 *Under the conditions of Theorem 7.1 and under local alternative (7.26), the R-estimator $\widehat{\boldsymbol{\beta}}_n$ has asymptotic normal distribution*

$$n^{1/2}(\widehat{\boldsymbol{\beta}}_n - (\mathbf{Q} + \mathbf{V})^{-1}\mathbf{Q}\,\boldsymbol{\beta}_n) \xrightarrow{\mathcal{D}} \mathcal{N}_p \left(\mathbf{0}, (\mathbf{Q} + \mathbf{V})^{-1}A^2(\varphi) \right). \qquad (7.29)$$

Notice that the local asymptotic bias cannot be controlled by the choice of the score-generating function φ; this choice can only influence the asymptotic variance factor of the estimator. The magnitude of the bias fully depends on the precision of the measurements, namely on the matrix $\mathbf{V}$. The measurement errors in the responses Y_{ni} affect only the asymptotic variance, not the bias. The result is entirely non-parametric, valid for a class of distributions of model errors, demanding only the finite (and positive) Fisher's information.

We can compare two measurement methods with the same regressors (random or non-random), with the respective limiting covariance matrices $\mathbf{V}_1$, $\mathbf{V}_2$. Comparing the biases in (7.28) for $\mathbf{V}_1$ and $\mathbf{V}_2$, the first method is considered being more precise than the second one if the matrix $(\mathbf{V}_2+\mathbf{Q})^{-1} \prec (\mathbf{V}_1+\mathbf{Q})^{-1}$; in other words, if $\mathbf{Q}^{-1}\mathbf{V}_1 \prec \mathbf{Q}^{-1}\mathbf{V}_2$ where ordering $\mathbf{A} \prec \mathbf{B}$ means that matrix $\mathbf{B} - \mathbf{A}$ is positive definite.

With the function `rfit` and the following code we can illustrate the bias for the contaminated engel data.

```
> library(quantreg)
> data(engel)
> library(Rfit)
> (r.fit <- rfit(foodexp~income, data=engel))
Call:
rfit.default(formula = foodexp ~ income, data = engel)

Coefficients:
(Intercept)        income
103.6453126    0.5377772


###
> foodexp.con<-engel[,2]+runif(235,-10,10)
> income.con <-engel[,1]+runif(235,-10,10)
> (r.fit.con <- rfit(foodexp.con~income.con))
Call:
rfit.default(formula = foodexp.con ~ income.con)

Coefficients:
(Intercept)  income.con
102.0480779    0.5402829
```

7.5   Problems and complements

7.1 Let f_δ be the density of $Z_\delta = X + \sqrt{\delta}V$, i.e., X is affected by an additive measurement error $\sqrt{\delta}\,V$. Assume that X and V are independent, X has continuous symmetric density f_0 and V has strongly unimodal density satisfying $EV = 0$, $EV^2 = 1$, $EV^4 < \infty$, otherwise unknown. If, moreover,

f_0 has differentiable and integrable derivatives up to order 5, then we have the following expansion of f_δ under $\delta \downarrow 0$:

$$f_\delta(z) = f_0(x + \sqrt{\delta}V) = f_0(x) + \frac{\delta}{2}\frac{d^2}{dz^2}f_0(x) + \frac{\delta^2}{4!}\frac{d^4}{dz^4}f_0(x)E(V^4) + o(\delta^2).$$

7.2 Let X and Y have twice differentiable densities f_0 and g_0, and f_δ and g_δ be the respective densities of $Z_{\delta 1} = X + \sqrt{\delta}V$ and $Z_{\delta 2} = Y + \sqrt{\delta}V$, where V is independent of X and Y and $EV = 0$, $EV^2 = 1$. If $X \ll Y$ and

$$\lim_{z \to \infty} \frac{d}{dz}\left[f_0(z)\log\left(\frac{f_0(z)}{g_0(z)}\right)\right] = 0,$$

then

$$\frac{d}{d\delta}D\left(f_\delta, g_\delta\right)\Big|_{\delta=0+} = -\frac{1}{2}E_f\left\{(\nabla \log f(Z) - \nabla \log g(Z))^2\right\}$$

where

$$D(f,g) = \int \log\left(\frac{f}{g}\right)f(z)dz$$

is the Kullback-Leibler divergence of f, g and $\nabla \log f(x) = \frac{f'(x)}{f(x)}$. (Guo (2009).)

7.3 Generate samples of the linear regression model with different distribution of errors and contaminations and apply the function `rfit` to these data.

7.4 Apply the functions `rfit` on the dataset used in Chapter 4 (the Hertzsprung-Russell diagram of the Star Cluster `starsCYG`, `father.son` dataset) for other score functions and compare the obtained results.

7.5 Apply the function `rank.test.meas` from Appendix A to testing of the regression coefficient on the datasets of Chapter 4 and compare the obtained results.

Appendix A

Authors' own procedures in R

also available online at: http://robust.tul.cz/

```
huber2 <- function (y, k = 1.345, initmu = median(y),
s = mad(y), iters = FALSE, tol = 1e-08)
{
require(robustbase)
y <- y[!is.na(y)]
n <- length(y)
mu <- as.numeric(initmu)
s <- as.numeric(s)
Niter <- 300
Converged <- FALSE
if (s == 0)
stop("cannot estimate scale: s is zero for this sample")
for (i in 0:Niter) {
if (iters) s <- mad(y,mu)   #s ignored, MAD of current residuals
yy <- Mpsi((y - mu)/s, cc = k, psi="huber")*s
mu1 <- sum(yy)/n
mu <- mu + mu1
if (abs(mu1) < tol * s ) {Converged <- TRUE; Niter <- i; break}
}
list(mu = mu, s = s, Niter = Niter, Converged = Converged)
}


cauchy <- function (y, initmu = median(y),
s = mad(y), iters = FALSE, tol = 1e-08)
{
y <- y[!is.na(y)]
n <- length(y)
mu <- as.numeric(initmu)
s <- as.numeric(s)
```

```
Niter <- 300
Converged <- FALSE
if (s == 0)
stop("cannot estimate scale: s is zero for this sample")
for (i in 0:Niter) {
if (iters) s <- mad(y,mu)  #s ignored, MAD of current residuals
x <- (y - mu)/s
yy <- 2 * x/(1 + x^2) * s
mu1 <- sum(yy)/n
mu <- mu + mu1
if (abs(mu1) < tol * s ) {Converged <- TRUE; Niter <- i; break}
}
list(mu = mu, s = s, Niter = Niter, Converged = Converged)
}


biweight <- function (y, k = 4.685061, initmu = median(y),
s = mad(y), iters = FALSE, tol = 1e-08)
{
require(robustbase)
y <- y[!is.na(y)]
n <- length(y)
mu <- as.numeric(initmu)
s <- as.numeric(s)
Niter <- 300
Converged <- FALSE
if (s == 0)
stop("cannot estimate scale: s is zero for this sample")
for (i in 0:Niter) {
if (iters) s <- mad(y,mu)  #s ignored, MAD of current residuals
yy <- Mpsi((y - mu)/s, cc = k, psi="bisquare") * s
mu1 <- sum(yy)/n
mu <- mu + mu1
if (abs(mu1) < tol * s ) {Converged <- TRUE; Niter <- i; break}
}
list(mu = mu, s = s, Niter = Niter, Converged = Converged)
}


hampel <- function (y, a = 2, b = 4, c = 8, initmu = median(y),
s = mad(y), iters = FALSE, tol = 1e-08)
{
require(robustbase)
y  <- y[!is.na(y)]
n  <- length(y)
```

```
mu <- as.numeric(initmu)
s  <- as.numeric(s)
Niter <- 300
Converged <- FALSE
if (s == 0)
stop("cannot estimate scale: s is zero for this sample")
for (i in 0:Niter) {
if (iters) s <- mad(y,mu) #s ignored, MAD of current residuals
yy  <- Mpsi((y - mu)/s, cc = c(a, b, c), psi="hampel") * s
mu1 <- sum(yy)/n
mu  <- mu + mu1
if (abs(mu1) < tol * s ) {Converged <- TRUE; Niter <- i; break}
}
list(mu = mu, s = s, Niter = Niter, Converged = Converged)
}


skipped.mean <- function (y, k = 2, initmu = median(y),
s = mad(y), iters = FALSE, tol = 1e-08)
{
y <- y[!is.na(y)]
n <- length(y)
mu <- as.numeric(initmu)
s <- as.numeric(s)
Niter <- 300
Converged <- FALSE
if (s == 0)
stop("cannot estimate scale: s is zero for this sample")
for (i in 0:Niter) {
if (iters) s <- mad(y,mu)  #s ignored, MAD of current residuals
x <- (y - mu)/s
yy <- ifelse(abs(x) > k, 0, x) * s
mu1 <- sum(yy)/n
mu <- mu + mu1
if (abs(mu1) < tol * s ) {Converged <- TRUE; Niter <- i; break}
}
list(mu = mu, s = s, Niter = Niter, Converged = Converged)
}


winsorized.mean<-function(x, trim=0)
{
x <-sort(x[!is.na(x)])
n<-length(x)
if ((trim < 0) | (trim>0.5) )
```

```
stop("cannot estimate: alpha<0 or alpha>0.5")
ntr<-trunc(trim*n)
(ntr*x[ntr+1]+sum(x[(ntr+1):(n-ntr)])+ntr*x[n-ntr])/n
}



hodges.lehmann <- function(x)
{
x <- x[!is.na(x)]
diffs <- outer(x, x, "+")
diffs <- diffs[!upper.tri(diffs)]
median(diffs)/2
}



tls.KB<-function(formula, data,  alpha=0.05)
{
require(quantreg)
resid1 <-residuals(rq(formula,data,tau=alpha))
resid2 <-residuals(rq(formula,data,tau=1-alpha))
c1 <- c(resid1 >= 0)
c2 <- c(resid2 <= 0)
coefficients(lm(formula,data[c(c1 & c2),]))
}



rfit0 <- function(formula, data=NULL, tau=0.5)
{
require(Rfit)
n <- dim(model.frame(formula,data))[1]
my.phi0 <- function(u, param = tau)
{
s1 = param[1]
c1 = ceiling(s1*n)
ifelse(u < c1/(n+1), s1 - 1, ifelse(u > c1/(n+1), s1,
c1-1-s1*(n-1)))
}
my.Dphi0 <- function(u, param = tau)
{
s1 = param[1]
ifelse(u < s1, 0, ifelse(u > s1, 0, 1))
}
myscores0 <- new("scores", phi = my.phi0, Dphi = my.Dphi0,
param = tau)
rfit(formula=formula, data=data,scores=myscores0)
}
```

```
disp0 <- function (beta, x, y, tau=0.5)
{
require(Rfit)
my.phi0 <- function(u, param)
{
s1 = param[1]
c1 = ceiling(s1*n)
ifelse(u < c1/(n+1), s1 - 1, ifelse(u > c1/(n+1), s1,
c1-1-s1*(n-1)))
}
x <- as.matrix(x)
e <- y - x %*% beta
r <- rank(e, ties.method = "first")/(length(e) + 1)
my.phi0(r,param=tau) %*% e
}


"LADLS"<-function(ydata, xdata, resultls, resultl1, int=T,
  kern=F)
{
 n <- length(ydata)
 xdata <- as.matrix(xdata)
 p <-length(xdata[1,])
 if(!int)
   p<-p-1
 if(missing(resultl1))
    resultl1 <- rq(ydata~xdata,tau=0.5,ci=F)
 if(missing(resultls))
    resultls <- lm(ydata~xdata)
 if (kern) sn<-s.kernel(xdata,ydata,sqrt(3),int=int)
 else sn<-s.histog(xdata,ydata,int=int)
 absresl1 <- abs(resultl1[[2]])
 e10 <- 1/(sn * (n - p -1)) * sum(absresl1)
 sigma02 <- (1/((sn^2) * (n - p - 1))) * sum(resultl1[[2]]
            * resultl1[[2]])
 if(int)
   xdata<-cbind(rep(1,n),xdata)
 if(e10 >= 0.5) {
   delta0 <- 1
   tn <- resultl1[[1]]
 }
 if(e10 < 0.5 && e10 >= 2 * sigma02) {
   delta0 <- 0
   tn <- resultls[[1]]
 }
```

```
if(e10 < 0.5 && e10 < 2 * sigma02) {
 delta0 <- (4 * sigma02 - 2 * e10)/(4 * sigma02 - 4 * e10 + 1)
 tn0 <- delta0 * resultl1[[1]] + (1 - delta0) * resultls[[1]]
 tn<-nlminb(start=tn0, objective=my2.fcn, matx=xdata,
                    vecy=ydata,delta=delta0)$parameters
}
tn <- as.vector(tn)
if(int)
    xlabel <- c("Intercept", paste("X", 1:p, sep= ""))
else xlabel <- paste("X", 1:(p+1), sep = "")
names(tn)<-xlabel
return(coef=tn)
}


"my.fcn" <- function(t,matx,vecy,delta,gama,sn,k)
         { (1-delta)*2/gama* sum(((((vecy-matx%*%t)/sn)^2)/2)
         [c(compare(rep(k,length(vecy))),
           (vecy-matx%*%t)/sn))>=0]) +
         (1-delta)*2/gama*
         sum((k*abs((vecy-matx%*%t)/sn)^(-k*k/2))
         [c(compare(rep(k,length(vecy))),
           (vecy-matx%*%t)/sn))<0]) +
         delta*sum(abs(vecy-matx%*%t)/sn)
         }


"my2.fcn" <- function(t,matx,vecy,delta)
          { (1-delta)* sum(((vecy-matx%*%t)^2)) +
            delta*sum(abs(vecy-matx%*%t))
          }


 "s.histog"<-function(xdata,ydata,int=T)
{
 nun <- (((1/8)^(2/5)) * (9/2)^(1/5) * (length(ydata))^(-1/5))
 cxdata <- scale(xdata, center = T, scale = F)
 quantplus <- rq(ydata~cxdata,  tau=(0.5 + nun),ci=F)$coef[1]
 quantminus <- rq(ydata~cxdata, tau=(0.5 - nun),ci=F)$coef[1]
 sn <- max((quantplus-quantminus)/(2 * nun),0.0001)
 return(sn)
}


"s.kernel"<-function(xdata,ydata,b,int=T)
{
 nun <- (((1/8)^(2/5)) * (9/2)^(1/5) * (length(ydata))^(-1/5))
 sol<-rq(ydata~xdata,ci=F)$sol[c(1,3),]
 sol<-sol[,abs((0.5-sol[1,])/nun)<b]
```

```
 if (sol[1,1]>0){
                xl<-0.5-b*nun
                xl<-c(xl,rq(ydata~xdata,tau=xl,ci=F)$coef[1])
                sol<-cbind(xl,sol)
 }
 if (sol[1,length(sol[1,])]<1){
        xu<-0.5+b*nun
        xu<-c(xu,rq(ydata~xdata,int=int,tau=xu,ci=F)$coef[1])
        sol<-cbind(sol,xu)
 }
 nsol<-length(sol[1,])
 kernel<--1.5/b^3*(0.5*sol[1,]/nun-sol[1,]^2/(2*nun))
 snker<-0
 for(i in 1:(nsol-1))
  {snker<-snker + sol[2,i]*(kernel[i+1]-kernel[i])}
 snker<-max(snker/nun^2,0.0001)
 return(snker)
}


rank.test.meas<-
function(x, y, score = "wilcoxon")
{
n<-length(x[,1])
xp<-apply(x,2,mean)
x2<-x-matrix(rep(xp,n),nrow=n,byrow=T)
Q=t(x2)%*%x2/n
por<-rank(y)
if(score == "wilcoxon") {
skor<-por/(n+1)-0.5
A2<-12
}
 else if(score == "normal") {
  skor<-qnorm(por/(n+1))
  A2<-1
}
else if(score == "sign") {
  skor<-0.5*sign(por/(n+1)-0.5)
  A2<-1
}
sn<-n^(-0.5)*apply(x2*skor,2,sum)
t(sn)%*%solve(Q)%*%sn*A2
}
```

Bibliography

[1] R.J. Adcock (1877). Note on the method of least squares. *The Analyst 4*, 183–184.

[2] J.G. Adrover and V.J. Yohai (2002). Projection estimates of multivariate location. *Ann. Statist.* 20, 1760–1781.

[3] M.G. Akritas and M.A. Bershady (1996). Linear regression for astronomical data with measurement errors and intrinsic scatter. *Astrophysical Journal* 470, 706–728.

[4] P. Amrhein (1995). An example of a two-sided Wilcoxon signed rank test which is not unbiased. *Ann. Inst. Statist. Math.* 47, 167–170.

[5] Analytical Methods Committee (1989). Robust statistics: How not to reject outliers. *The Analyst 114*, 1693–1702.

[6] D.F. Andrews, P.J. Bickel, F.R. Hampel, P.J. Huber, W.H. Rogers, and J.W. Tukey (1972). *Robust Estimates of Location. Survey and Advances.* Princeton University Press, Princeton, NJ.

[7] J. Antoch and J.Á. Víšek (eds.) (1992). *Computational Aspects of Model Choice.* Physica-Verlag, Heidelberg.

[8] J. Antoch, H. Ekblom and J.Á. Víšek (1998). *Robust Estimation in Linear Model.* XploRe Macros: http://www.quantlet.de/codes/rob/ROB.html.

[9] O. Arias, K.F. Hallock, W. Sosa-Escudero (2001). Individual heterogeneity in the returns to schooling: Instrumental variables quantile regression using twins data. *Empirical Economics* 26, 7–40.

[10] T.S. Arthanari and Y. Dodge (1981). *Mathematical Programming in Statistics.* J. Wiley, 1993 in Wiley Classic Library.

[11] R.R. Bahadur (1967). Rates of convergence of estimators and test statistics. *Ann. Math. Statist.* 38, 303–324.

[12] A.D. Barbour and P. Hall (1984). On the rate of Poisson convergence. *Math. Proc. Cambridge Philos. Soc.* 95, 473–480.

[13] V.D. Barnett (1966). Evaluation of the maximum likelihood estimator where the likelihood equation has multiple roots. *Biometrika* 53, 152–166.

[14] V.D. Barnett and T. Lewis (1994). *Outliers in Statistical Data,* 3rd ed. John Wiley & Sons, Chichester, U.K.

[15] G.W. Bassett (1991). Equivariant, monotonic, 50% breakdown estimators. *Amer. Statist.* 45, 135–137.

[16] D.A. Belsley, E. Kuh and R.E. Welsch (1980). *Regression Diagnostics: Identifying Influential Data and Sources of Collinearity.* John Wiley & Sons, New York.

[17] A. Bhattacharyya, (1943). On a measure of divergence between two statistical populations defined by their probability distribution. *Bull. Calcutta Math. Soc.* 35, 99–109.

[18] P.J. Bickel (1969). A distribution free version of the Smirnov two sample test in the p-variate case. *Ann. Statist.* 40, 1–23.

[19] P.J. Bickel (1975). One-step Huber estimates in the linear model. *Ann. Statist.* 1, 597–616.

[20] P.J. Bickel and E.L. Lehmann (1979). Descriptive statistics for nonparametric model. IV. Spread. *Contributions to Statistics: Jaroslav Hájek Memorial Volume* (J. Jurečková, ed.), pp. 33–40. Academia, Prague and Reidel, Dordrecht.

[21] P.J. Bickel, C.A.J. Klaassen, Y. Ritov and J.A. Wellner (1993). *Efficient and Adaptive Estimation for Semiparametric Models.* John Hopkins University Press, Baltimore.

[22] P. Billingsley (1998). *Convergence of Probability Measures,* 2nd ed. John Wiley & Sons, New York.

[23] G. Blom (1956). On linear estimates with nearly minimum variance. *Arkiv für Mathematik* 3, 365–369.

[24] P. Bloomfield and W.L. Steiger (1983). *Least Absolute Deviations. Theory, Applications and Algorithms.* Birkhäuser, Boston.

[25] R.M. Blumenthal and R.K. Getoor (1968). *Markov Processes and Potential Theory.* Academic Press, New York.

[26] R.J. Boskovich (1757). De literaria expeditione per pontificiam ditionem et synopsis amplioris operis... *Bononiensi Scientiarum et Artum Instituto atque Academia Commentarii* 4, 353–396.

[27] G.E.P. Box (1953). Non-normality and tests of variance. *Biometrika* 40, 318–335.

[28] G.E.P. Box and S.L. Anderson (1955). Permutation theory in the derivation of robust criteria and the study of departures from assumption. *J. Royal Statist. Soc.* B 17, 1–34.

[29] A.C. Brandwein and W.E. Strawderman (1991). Generalizations of James-Stein estimators under spherical symmetry. *Ann. Statist.* 19, 1639–1650.

[30] B.M. Brown (1982). Cramér-von Mises distributions and permutation tests. *Biometrika* 69, 619–624.

[31] B.M. Brown (1983). Statistical uses of the spatial median *J. Royal Statist. Soc.* B 45, 25–30.

[32] B.M. Brown (1988). Spatial median. *Encyclopedia of Statistical Sciences*, Vol. 8 (S. Kotz, N.L. Johnson and C.B. Read, eds.), pp. 574–575. John Wiley & Sons, New York.

[33] K.A. Brownlee (1960). *Statistical Theory and Methodology in Science and Engineering*. John Wiley & Sons, New York, 491–500.

[34] H. Bunke and O. Bunke (eds.) (1986). *Statistical Inference in Linear Models*. John Wiley & Sons, Chichester, U.K.

[35] R.J. Carroll and D. Ruppert (1988). *Transformations and Weighting in Regression*. Chapman & Hall, London.

[36] R.J. Carroll, D. Ruppert, L.D. Stefanski and C.M. Crainiceanu (2006). *Measurement Error in Nonlinear Models. A Modern Perspective* (Second Edition). Chapman & Hall/CRC.

[37] R.J. Carroll, J.D. Maca, D. Ruppert (1999). Nonparametric regression in the presence of measurement error. *Biometrika* 86, 541–554.

[38] R.J. Carroll, A. Delaigle and P. Hall (2007). Non-parametric regression estimation from data contaminated by a mixture of Berkson and classical errors. *Journal of Royal Statistical Society* B 69, part 5, 859–878.

[39] D. Cellier and D. Fourdrinier (1995). Shrinkage estimators under spherical symmetry for the general linear model. *J. Multiv. Anal.* 52, 338–351.

[40] D. Cellier, D. Fourdrinier and C. Robert (1989). Robust shrinkage estimators of the location parameter for elliptically symmetric distributions. *J. Multiv. Anal.* 29, 39–52.

[41] B. Chakraborty , P. Chaudhuri and H. Oja (1998). Operating transformation retransformation on spatial median and angle test. *Statistica Sinica* 8, 767–784.

[42] B. Chakraborty (2001). On affine equivariant multivariate quantiles. *Annals of the Institute of Statistical Mathematics* 53, 380–403.

[43] S. Chaterjee and A.S. Hadi (1988). *Sensitivity Analysis in Linear Regression*. John Wiley & Sons, New York.

[44] P. Chaudhuri (1996). On a geometric notion of quantiles for multivariate data. *J. Amer. Statist. Assoc.* 91, 862–872.

[45] P. Chaudhuri and D. Sengupta (1993). Sign tests in multi-dimension: Inference based on the geometry of the data cloud. *J. Amer. Statist. Assoc. 88*, 1363–1370.

[46] C.L. Cheng and J.W. van Ness (1999). *Statistical Regression with Measurement Error*. Arnold, London.

[47] H. Chernoff and H. Teicher (1958). A central limit theorem for sums of interchangeable random variables. *Ann. Math. Statist.* 29, 118–130.

[48] K.I. Choi and J. Marden (1997). An Approach to multivariate rank tests in multivariate analysis of variance. *J. Amer. Statist. Assoc.* 92, 1582–1990.

[49] Y.S. Chow and K.F. Yu (1981). The performance of a sequential procedure for the estimation of the mean. *Ann. Statist.* 9, 184–188.

[50] K.L. Chung (1982). *Lectures from Markov Processes to Brownian Motion.* Springer-Verlag, Berlin.

[51] B.R. Clarke (2018). *Robustness Theory and Application.* J. Wiley.

[52] J.R. Collins (1982). Robust M-estimators of location vectors. *J. Multivar. Analysis* 12, 480–492.

[53] J.R. Collins (1986). Maximum asymptotic variances of trimmed means under asymmetric contaminations. *Ann. Statist.* 14, 348–354.

[54] J.R. Collins (1991). Maximum asymptotic biases and variances of symmetrized interquantile ranges under asymmetric contamination. *Statistics* 22, 379–402.

[55] J.R. Collins (1999). Robust M-estimators of scale: Minimax bias versus maximal variance. *Canad. J. Statist.* 27, 81–96.

[56] J.R. Collins (2000). Robustness comparisons of some classes of location parameter estimators. *Ann. Inst. Statist. Math.* 52, 351–366.

[57] J.R. Collins (2003). Bias-robust L-estimators of a scale parameter. *Statistics* 37, 287–303.

[58] J.R. Collins and B.Wu (1998). Comparisons of asymptotic biases and variances of M-estimators of scale under asymmetric contamination. *Commun. Statist. Theory Meth.* 27, 1791–1810.

[59] R.D. Cook and S. Weisberg (1982). *Residuals and Influence in Regression.* Chapman & Hall, London.

[60] I. Csiszár (1967). Information-type measures of difference of probability distributions and indirect observations. *Studia Sci. Math. Hungar.* 2, 299–318.

[61] M. Csörgő, and P. Révész (1978). Strong approximation of the quantile process. *Ann. Statist.* 6, 882–894.

[62] M. Csörgő, and P. Révész (1981). *Strong Approximations in Probability and Statistics.* Akadémiai Kiadó, Budapest.

[63] P.L. Davies (1987). Asymptotic behavior of S-estimates of multivariate location parameters and dispersion matrices. *Ann. Statist.* 15, 1269–1292.

[64] P.L. Davies (1990). The asymptotics of S-estimators in the linear regression model. *Ann. Statist.* 18, 1651–1675.

[65] P.L. Davies (1992). The asymptotics of Rousseeuw's minimum volume ellipsoid estimator. *Ann. Statist.* 20, 1828–1843.

[66] P.L. Davies and U. Gather (2004). Robust statistics. *Handbook of Computational Statistics* (J.E. Gentle, W. Härdle, Y. Mori, eds.), pp. 655–695. Springer-Verlag, Berlin.

[67] A.P. Dempster (1969). *Elements of Continuous Multivariate Analysis.* Addison-Wesley, New York.

[68] S.J. Devlin, R. Gnanadesikan and J.R. Kettenring (1976). Some multivariate applications of elliptical distributions. *Essays in Probability and Statistics* (S. Ikeda, ed.), pp. 365–395. Shinko Tsusho, Tokyo.

[69] T. de Wet and J.H. Venter (1973). Asymptotic distributions of quadratic forms with application to test of fit. *Ann. Statist.* 31, 276–295.

[70] R.L. Dobrushin (1970). Describing a system of random variables by conditional distributions. *Theor. Prob. Appl.* 15, 458–486.

[71] Y. Dodge (1984). Robust estimation of regression coefficient by minimizing a convex combination of least squares and least absolute deviations. *Comp. Statist. Quarterly* 1, 139–153.

[72] Y. Dodge (1996). The guinea pig of multiple regression. *Robust Statistics, Data Analysis, and Computer Intensive Methods*, (H. Rieder, ed.), Lecture Notes in Statistics 109, pp. 91–118. Springer-Verlag, New York.

[73] Y. Dodge, J. Antoch and J. Jurečková (1991). Adaptive combination of least squares and least absolute deviations estimators: Computational aspects. *Comp. Stat. Data Anal.* 12, 87–99.

[74] Y. Dodge and J. Jurečková (1987). Adaptive combination of least squares and least absolute deviations estimators. *Statistical Data Analysis Based on L_1 Norm and Related Methods* (Y. Dodge, ed.), pp. 275–284.

[75] Y. Dodge and J. Jurečková (1988). Adaptive combination of M-estimator and L_1-estimator in the linear model. *Optimal Design and Analysis of Experiments* (Y. Dodge, V. V. Fedorov, H. P. Wyn, eds.), pp. 167–176. North-Holland, Amsterdam.

[76] Y. Dodge and J. Jurečková (1991). Flexible L-estimation in the linear model. *Comp. Stat. Data Anal.* 12, 211–220.

[77] Y. Dodge and J. Jurečková (1995). Estimation of quantile density function based on regression quantiles. *Statistics & Probability Letters* 23, 73–78.

[78] Y. Dodge and J. Jurečková (2000). *Adaptive Regression.* Springer-Verlag, New York.

[79] Y. Dodge and P.T. Lindstrom (1981). An alternative to least square estimation when dealing with contaminated data. *Technical Report* No 79, Oregon State University, Corvalis.

[80] D.L. Donoho (1982). *Breakdown properties of multivariate location estimators.* Ph.D. Thesis, Department of Statistics, Harvard University, Harvard, MA.

[81] D.L. Donoho and M. Gasko (1992). Breakdown properties of location estimates based on half space depth and projection outlyingness. *Ann. Statist.* 20, 1803–1827.

[82] D.L. Donoho and P.J. Huber (1983). The notion of breakdown point. *A Festschrift for Erich Lehmann* (P.J. Bickel, K.A. Doksum and J.L. Hodges, eds.), pp. 157–184. Wadsworth, CA.

[83] J.L. Doob (1984). *Classical Potential Theory and Its Probabilistic Counterpart.* Springer.

[84] N.R. Draper and H. Smith (1988). *Applied Regression Analysis*, 3rd ed. John Wiley & Sons, New York.

[85] S.P. Ellis (1998). Instability of least squares, least absolute deviation and least median of squares linear regression. *Statist. Science* 13, 337–350.

[86] M. Fabian, P. Habala, P. Hájek, V.M. Santalucía, J. Pelant and V. Zízler (2001). *Functional Analysis and Infinite-Dimensional Geometry.* Springer, New York.

[87] M. Falk (1986). On the estimation of the quantile density function. *Statist. Probab. Lett.* 4, 69–73.

[88] J. Fan and Y.K. Truong (1993). Nonparametric regression estimation involving errors-in-variables. *Ann. Statist.* 21, 23–37.

[89] L.T. Fernholz (1983). *von Mises Calculus for Statistical Functionals.* Lecture Notes in Statistics 19, Springer-Verlag, New York.

[90] C.A. Field and E.M. Ronchetti (1990). *Small Sample Asymptotics.* IMS Lecture Notes 13, IMS, Hayward, CA.

[91] R.A. Fisher (1922). On the mathematical foundations of theoretical statistics. *Phil. Trans. Roy. Soc. London* A 222, 309–368.

[92] D. Fourdrinier, E. Marchand and E. Strawderman (2004). On the inevitability of a paradox in shrinkage estimation for scale mixture of normals. *J. Statist. Plann. Infer.* 121, 37–51.

[93] D. Fourdrinier and E. Strawderman (1996). A paradox concerning shrinkage estimators: Should a known scale parameter be replaced by an estimated value in the shrinkage factor? *J. Multiv. Anal.* 59, 109–140.

[94] J.H. Friedman and L.C. Rafsky (1979). Multivariate generalizations of the Wald-Wolfowitz and Smirnov two-sample tests. *Ann. Statist.* 7, 697–717.

[95] J.C. Fu (1975). The rate of convergence of consistent point estimators. *Ann. Statist.* 3, 234–240.

[96] J.C. Fu (1980). Large deviations, Bahadur efficiency, and the rate of convergence of linear functions of order statistics. *Bull. Inst. Math. Acad. Sinica* 8, 15–37.

[97] W.A. Fuller (1987). *Measurement Error Models.* J. Wiley, New York.

[98] M. Ghosh and N. Mukhopadhyay (1979). Sequential point estimation of the mean when the distribution is unspecified. *Comm. Statist.* A 8, 637–652.

[99] A.L. Gibbs and F.E. Su (2002). On choosing and bounding probability metrics. *Intern. Statist. Rev.* 70, 419–435.

[100] S.D. Grose and M.L. King (1991). The locally unbiased two-sided Durbin-Watson test. *Economics Letters* 35, 401–407.

[101] D. Guo (2009). Relative entropy and score function: New information-estimation relationships through arbitrary additive perturbation. *ISIT 2009, Seoul, Korea*, 814–818.

[102] C. Gutenbrunner (1986). Zur Asymptotik von Regression Quantil Prozessen und daraus abgeleiten Statistiken. Ph.D. Thesis, Universität Freiburg, Germany.

[103] C. Gutenbrunner and J. Jurečková (1992). Regression rank scores and regression quantiles. *Ann. Statist.* 20, 305–330.

[104] C. Gutenbrunner, J. Jurečková, R. Koenker and S. Portnoy (1993). Tests of linear hypotheses based on regression rank scores. *J. Nonpar. Statist.* 2, 307–331.

[105] J. Hájek (1968). Asymptotic normality of simple linear rank statistics under alternatives. *Ann. Math. Statist.* 39, 325–346.

[106] J. Hájek (1969). *A Course in Nonparametric Statistics.* Holden-Day, San Francisco.

[107] J. Hájek, Z. Šidák, P.K.Sen (1999). *Theory of Rank Tests.* Academic Press.

[108] P.J. Hall (1981). Large sample properties of Jaeckel's adaptive trimmed mean. *Ann. Inst. Statist. Math.* 33A, 449–462.

[109] P.J. Hall and N. Tajvidi (2002). Permutation tests for equality of distributions in high-dimensional settings. *Biometrika* 89, 359–374.

[110] M. Hallin and J. Jurečková (1999). Optimal tests for autoregressive models based on autoregression rank scores. *Ann. Statist.* 27, 1385–1414.

[111] M. Hallin and D. Pandaveine (2002). Optimal tests for multivariate location based on interdirections and pseudo-Mahalanobis ranks. *Ann. Statist.* 30, 1103–1133.

[112] M. Hallin, D. Paindaveine and T. Verdebout (2010). Testing for common principal components under heterokurticity. *Journal of Nonparametric Statistics* 22, 879–895.

[113] F.R. Hampel (1968). *Contribution to the Theory of Robust Estimators.* Ph.D. Thesis. University of California, Berkeley, CA.

[114] F.R. Hampel (1971). A general qualitative definition of robustness. *Ann. Math. Statist.* 42, 1887–1896.

[115] F.R. Hampel (1973). Some small sample asymptotics. *Proceedings of the Prague Symposium on Asymptotic Statistics*, Vol. 2 (J. Hájek, ed.), pp. 109–126. Charles University, Prague.

[116] F.R. Hampel (1974). The influence curve and its role in robust estimation. *J. Amer. Statist. Assoc.* 69, 383–393.

[117] F.R. Hampel (1975). Beyond location parameters: Robust concepts and methods (with discussion). *Bull. Internat. Statist. Inst.* 46, 375–391.

[118] F.R. Hampel, P.J. Rousseeuw, E.M. Ronchetti, and W. Stahel (1986, 2005). *Robust Statistics – The Approach Based on Influence Functions.* John Wiley & Sons, New York. Russian translation 1989.

[119] F. Harrell and C. Davis (1982). A new distribution-free quantile estimator. *Biometrika* 69, 636–640.

[120] J. Hausman (2001). Mismeasured variables in econometric analysis: Problems from the right and problems from the left. *J. Econ. Perspect.* 15/4, 57–67.

[121] X. He and H. Liang (2000). Quantile regression estimate for a class of linear and partially linear errors-in-variables models. *Statistica Sinica* 10, 129–140.

[122] T.P. Hettmansperger (1985). *Statistical Inference Based on Ranks.* John Wiley & Sons, New York.

[123] P. Harremoës and P.S. Ruzankin (2004). Rate of convergence to Poisson law in terms of information divergence. *Trans. IEEE Inform. Theory* 50, 2145–2149.

[124] D.M. Hawkins and D. Olive (1999). Applications and algorithms for least trimmed sum of absolute deviations, regression. *Comp. Statist. Data Anal.* 32, 119–134.

[125] X. He, J. Jurečková, R. Koenker and S. Portnoy (1990). Tail behavior of regression estimators and their breakdown point. *Econometrica* 58, 1195–1214.

[126] L.L. Helms (1969). *Introduction to Potential Theory.* J. Wiley.

[127] X. He and D.G. Simpson (1993). Lower bounds for contamination bias: Globally minimax versus locally linear estimation. *Ann. Statist.* 21, 314–337.

[128] N. Henze (1988). A multivariate two-sample test based on the number of nearest neighbor type coincidences. *Ann. Statist.* 16, 772–783.

[129] T.P. Hettmansperger and J.W. McKean (2011). *Robust Nonparametric Statistical Methods*, Second Edition. Chapman& Hall-CRC

[130] T.P. Hettmansperger, J. Möttönen and H. Oja (1998). Affine invariant multivariate rank tests for several samples. *Statistica Sinica* **8**, 785–800.

[131] T.P. Hettmansperger and R.H. Randles (2002). A practical affine equivariant multivariate median. *Biometrika* 89, 851–860.

[132] T.P. Hettmansperger and S. Sheather (1992). A cautionary note on the method of least median squares. *Amer. Statist.* 46, 79–83.

[133] J.L. Hodges and E.L. Lehmann (1963). Estimation of location based on rank tests. *Ann. Math. Statist.* 34, 598–611.

[134] W. Hoeffding (1963). Probability inequalities for sums of bounded random variables. *J. Amer. Statist. Assoc.* 58, 13–29.

[135] W. Hoeffding (1973). On the centering of a simple linear rank statistic. *Ann. Statist.* 1, 54–66.

[136] O. Hössjer (1994). Rank-based estimates in the linear model with high breakdown point. *J. Amer. Statist. Assoc.* 89, 149–158.

[137] C. Huber-Carol, N. Balakrishnan, M.S. Nikulin, M. Mesbah (eds.) (2002). *Goodness-of-Fit Tests and Model Validity.* Birkhäuser, Boston.

[138] P.J. Huber (1964). Robust estimation of a location parameter. *Ann. Math. Statist.* 36, 73–101.

[139] P.J. Huber (1968). Robust confidence limits. *Z. Wahrsch. Verw. Geb.* 10, 269–278.

[140] P.J. Huber (1969). *Théorie de l'inférence de statistique robuste.* Presses de l'Université de Montréal, Montréal.

[141] P.J. Huber (1981). *Robust Statistics.* Wiley.

[142] P.J. Huber, E.M. Ronchetti (2009). *Robust Statistics*, J. Wiley, (2nd edition).

[143] W. Hyk and Z. Stojek (2013). Quantifying uncertainty of determination by standard additions and serial dilutions methods taking into account standard uncertainties in both axes. *Analytical Chemistry* 85, 5933–5939.

[144] D.R. Hyslop and Q.W. Imbens (2001). Bias from classical and other forms of measurement error. *Journal of Business & Economic Statistics* 19/4, 475–481.

[145] R. Ihaka, R. Gentleman (1996). R: A language for data analysis and graphics. *J. Computational Graphical Statistics* 5, 299–314.

[146] L.A. Jaeckel (1969). *Robust Estimates of Location.* Ph.D. Thesis, University of California, Berkeley, CA.

[147] L.A. Jaeckel (1971). Robust estimation of location: Symmetry and asymmetric contamination. *Ann. Math. Statist.* 42, 1020–1034.

[148] L.A. Jaeckel (1972). Estimating regression coefficients by minimizing the dispersion of residuals. *Ann. Statist.* 43, 1449–1458.

[149] W. James and C. Stein (1961). Estimation with quadratic loss. *Proceedings of the Fourth Berkeley Symposium on Mathematics Statistics and Probability*, Vol. 1, pp. 361–380. University of California Press, Berkeley, CA.

[150] P. Janssen, J. Jurečková and N. Veraverbeke (1985). Rate of convergence of one- and two-step M-estimators with applications to maximum likelihood and Pitman estimators. *Ann. Statist.* 13, 1222–1229.

[151] J. Jung (1955). On linear estimates defined by a continuous weight function. *Arkiv für Mathematik* 3, 199–209.

[152] J. Jung (1962). Approximation to the best linear estimates. *Contribution to Order Statistics* (A.E. Sarhan and B.G. Greenberg, eds.), pp. 28–33. John Wiley & Sons, New York.

[153] J. Jurečková (1977). Asymptotic relations of M-estimates and R-estimates in linear regression model. *Ann. Statist.* 5, 464–472.

[154] J. Jurečková (1980). Asymptotic representation of M-estimators of location. *Math. Operationsforsch. und Statistik, Ser. Statistics* 11, 61–73.

[155] J. Jurečková (1981). Tail-behavior of location estimators. *Ann. Statist.* 9, 578–585.

[156] J. Jurečková (2002). L_1 derivatives, score functions and tests. *Statistical Data Analysis Based on the L_1 Norm and Related Methods* (Y. Dodge, ed.), pp. 183–189. Birkhäuser, Basel.

[157] J. Jurečková (2012). Tail-behavior of estimators and of their one-step versions. *Journal de la Société Francaise de Statistique* 153, 44–51.

[158] J. Jurečková (2017). Regression quantile and average regression quiantile processes. *Analytical Methods in Statistics*, J. Antoch et al. (Eds.), Springer Proceedings in Mathematics and Statistics, pp. 53–62.

[159] J. Jurečková and J. Kalina (2012). Nonparametric multivariate rank tests and their unbiasedness. *Bernoulli* 18, 229–251.

[160] J. Jurečková and X. Milhaud (2003). Derivative of a density in the mean and statistical applications. *Festschrift for Constance van Eeden* (M. Moore et al., eds.). *IMS Lecture Notes* 42, pp. 217–232.

[161] J. Jurečková, M. Schindler, J. Picek (2017). Empirical regression quantile process with possible application to risk analysis. *arXiv:1710.06638*

[162] J. Jurečková and A.H. Welsh (1990). Asymptotic relations between L- and M-estimators in the linear model. *Ann. Inst. Statist. Math.* 42, 671–698.

[163] J. Jurečková, J. Picek and A.K.Md.E. Saleh (2010). Rank tests and regression rank scores tests in measurement error models. *Computational Statistics and Data Analysis* 54, 3108–3120.

[164] J. Jurečková and L.B. Klebanov (1997). Inadmissibility of robust estimators with respect to L_1 norm. (Y. Dodge ed.), *IMS Lecture Notes* 31, pp. 71–78.

[165] J. Jurečková and L.B. Klebanov (1998). Trimmed, Bayesian and admissible estimators. *Statist. Probab. Lett.* 42, 47–51.

[166] J. Jurečková and M. Malý (1995). The asymptotics for studentized k-step M-estimators of location. *Sequen. Anal.* 14, 229–245.

[167] J. Jurečková and X. Milhaud (1993). Shrinkage of maximum likelihood estimators of multivariate location. *Proceedings of 5th Prague Symposium on Asymptotic Statistics* (P. Mandl and M. Hušková , eds.), pp. 303–318. Physica-Verlag, Vienna.

[168] J. Jurečková, R. Koenker and S. Portnoy (2001). Tail behavior of the least squares estimator. *Statist. Probab. Lett.* 55, 377–384.

[169] J. Jurečková, H.L. Koul, R. Navrátil and J. Picek (2016). Behavior of R-estimators under measurement errors. *Bernoulli* 22, 1093–1112.

[170] J. Jurečková and R. Navrátil (2013). Rank tests under uncertainty: Regression and local heteroscedasticity. (R. Kruse et al., Eds.): *Synergies of Soft Computing and Statistics*, AISC 190, pp. 255–261, Springer.

[171] J. Jurečková and J. Picek (2011). Finite-sample behavior of robust estimators. *Recent Researches and Instrumentation, Measurement, Circuits and Systems* (S. Chen et al,. Eds.), pp. 15–20. ISBN: 978-960-474-282-0.

[172] J. Jurečková, J. Picek and P.K. Sen (2003). Goodness-of-fit test with nuisance regression and scale. *Metrika* 58, 235–258.

[173] J. Jurečková and S. Portnoy (1987). Asymptotics for one-step M-estimators in regression with application to combining efficiency and high breakdown point. *Commun. Statist. Theory Methods* A 16, 2187–2199.

[174] J. Jurečková and P.K. Sen (1982). M-estimators and L-estimators of location: Uniform integrability and asymptotic risk-efficient sequential versions. *Sequential Anal.* 1, 27–56.

[175] J. Jurečková and P.K. Sen (2006). Robust multivariate location estimation, admissibility and shrinkage phenomenon. *Statistics & Decisions* 24, 273–290.

[176] J. Jurečková and P.K. Sen (1990). Effect of the initial estimator on the asymptotic behavior of one-step M-estimator. *Ann. Inst. Statist. Math.* 42, 345–357.

[177] J. Jurečková and P.K. Sen (1994). Regression rank scores scale statistics and studentization in the linear model. *Proceedings of 5th Prague Conference on Asymptotic Statistics* (M. Hušková and P. Mandl, eds.), pp. 111–121. Physica-Verlag, Vienna.

[178] J. Jurečková and P.K. Sen (1996). *Robust Statistical Procedures: Asymptotics and Interrelations*. John Wiley & Sons, New York.

[179] J. Jurečková, P.K. Sen and J. Picek (2013). Methodological Tools in Robust and Nonparametric Statistics. Chapman & Hall/CRC Press, Boca Raton, London.

[180] A.M. Kagan, J.V. Linnik and C.R. Rao (1973). *Characterization Problems in Mathematical Statistics*. John Wiley & Sons, New York.

[181] B.C. Kelly (2007). Some Aspects of Measurement Error in Linear Regression of Astronomical Data. *The Astrophysical Journal* 665, 1489–1506.

[182] J.T. Kent and D.E. Tyler (1991). Redescending M-estimates of multivariate location and scatter. *Ann. Statist.* 19, 2102–2119.

[183] J.T. Kent, D.E. Tyler and Y. Vardi (1994). A curious likelihood identity for the multivariate t-distribution. *Comm. Statistics – Simulation Computation* 23, 441–453.

[184] J. Kim and D. Pollard (1990). Cube-root asymptotics. *Ann. Statist.* 18, 191–219.

[185] L.B. Klebanov (2005). *N-Distances and Their Applications.* Charles University in Prague, Karolinum Press.

[186] R. Koenker (2005). *Quantile Regression.* Cambridge University Press, Cambridge, U.K.

[187] R. Koenker and G. Bassett (1978). Regression quantiles. *Econometrica* 46, 33–50.

[188] R. Koenker and G. Bassett (1982). Robust tests of heteroscedasticity based on regression quantiles. *Econometrica* 50, 43–61.

[189] L. Kong and I. Mizera (2012). Quantile tomography: Using quantiles with multivariate data. *Statistica Sinica* 22, 1589–1610.

[190] I. Kontoyannis, P. Harremoës and O. Johnson (2005). Entropy and the law of small numbers. *IEEE Trans. Inform. Theory* 51, 466–472.

[191] H.L. Koul (2002). *Weighted Empirical Processes in Dynamic Nonlinear Models,* 2nd ed., Lecture Notes in Statistics 166, Springer, New York.

[192] T. Kubokawa, E. Marchand, W. E. Strawderman (2014). On improved shrinkage estimators for concave loss. CIRJE-F-936, 1–11.

[193] W. Krasker and R. Welsch (1982). Efficient bounded-influence regression estimation. *J. Amer. Statist. Assoc.* 77, 595–604.

[194] J.P. Lecoutre and P. Tassi (1987). *Statistique non parametrique et robustesse.* Economica, Paris.

[195] E.L. Lehmann (1983). *Theory of Point Estimators.* John Wiley & Sons, New York.

[196] E.L. Lehmann (1997). *Testing Statistical Hypotheses* (2nd edition). Springer, New York.

[197] E.L. Lehmann (1999). *Elements of Large Sample Theory.* Springer, New York.

[198] F. Liese and I. Vajda (1987). *Convex Statistical Distances.* Teubner, Leipzig, Germany.

[199] Yu. Yu. Linke (2017a). Asymptotic normality of one-step M-estimators based on non-identically distributed observations. *Statist. Probab. Lett.* 129, 216–221.

[200] Yu. Yu. Linke (2017b). Asymptotic properties of one-step weighted M-estimators and applications to some regression problems (in Russian). *Theory Probab. Appl.* 62, 468–498.

[201] Yu. Yu. Linke and I.S. Borisov (2017). Constructing initial estimators in one-step estimation procedures of nonlinear regression. *Statist. Probab. Lett.* 120, 87–94.

[202] Yu. Yu. Linke and A.I. Sakhanenko (2013). On asymptotics of the distribution of a two-step statistical estimator of a one-dimensional parameter (in Russian). Siberian Elektronic Matthematical Reports 10, 627–640.

[203] Yu. Yu. Linke and A.I. Sakhanenko (2013a). On asymptotics of the distributions of some two-step statistical estimators of a multidimensional parameter (in Russian). *Siberian Advances in Mathematics* 16/1, 89–120.

[204] Yu. Yu. Linke and A.I. Sakhanenko (2014). On conditions for asymptotic normality of Fisher's one-step estimators in one-parameter families of distributions (in Russian). Siberian Elektronic Matthematical Reports 11, 464–475.

[205] Yu. Yu. Linke (2016a). Refinement of Fisher's one-step estimates in the case of slowly converging initial estimators. *Theory Probab. Appl.* 60, 88–102.

[206] Yu. Yu. Linke and A.I. Sakhanenko (2016). Conditions of asymptotic normality of one-step M-estimators (in Russian). *Siberian Journal of Pure and Applied Mathematics* 16/4 , 46–64.

[207] R. Lie (1988). On a notion of simplicial depth. *Proc. Nat. Acad. Sci. USA* **85**, 1732–1734.

[208] R. Liu (1990). On a notion of data depth based on random simplices. *Ann. Statist.* **18**, 405–414.

[209] R. Liu and K. Singh (1993). A quality index based on data depth and multivariate rank tests. *J. Amer. Statist. Assoc.* **88**, 252–260.

[210] R.Y. Liu, J.M. Parelius and K. Singh (1999). Multivariate analysis by data depth: Descriptive statistics, graphics and inference (with discussion). *Ann. Statist.* 27, 783–858.

[211] H.P. Lopuhaä (1989). On the relation between S-estimators and M-estimators of multivariate location and covariance. *Ann. Statist.* 17, 1662–1683.

[212] H.P. Lopuhaä and P.J. Rousseeuw (1991). Breakdown properties of affine equivariant estimators of multivariate location and covariance matrices. *Ann. Statist.* 19, 229–248.

[213] J.-F. Maa, D.K. Pearl and R. Bartoszynski (1996). Reducing multidimensional two-sample data to one-dimensional interpoint comparisons. *Ann. Statist.* 24, 1069–1074.

[214] C.L. Mallows (1972). A note on asymptotic joint normality. *Ann. Math. Statist.* 43, 508–515.

[215] C. Mallows (1973). Influence functions. *National Bureau of Economic Research, Conference on Robust Regression*, Cambridge, MA.

[216] C. Mallows (1975). On some topics in robustness. Technical memorandum, Bell Telephone Laboratories, Murray Hill, NJ.

[217] A. Marazzi (1992). *Algorithms, Routines and S-Functions for Robust Statistics*. Chapman and Hall, New York.

[218] J.I. Marden (1998). Bivariate QQ plots and spider web plots. *Statistica Sinica* 8, 813–826.

[219] J.I. Marden and K. Choi (2005). Test of multivariate linear models using spatial concordances. *Journal of Nonparametric Statistics* 17, 167–183.

[220] R.A. Maronna (1976). Robust M-estimates of multivariate location and scatter. *Ann. Statist.* 4, 51–67.

[221] R.A. Maronna and V.J. Yohai (1981). Asymptotic behavior of general M-estimates for regression and scale with random carriers. *Z. Wahrscheinlichkeitstheorie und verw. Gebiete* 58, 7–20.

[222] R.A. Maronna and V.J. Yohai (1995). The behavior of the Stahel-Donoho robust multivariate estimator. *J. Amer. Statistc. Assoc.* 90, 330–341.

[223] R.A. Maronna, W.A. Stahel and V.J. Yohai (1992). Bias-robust estimates of multivariate scatter based on projections. *J. Multiv. Anal.* 21, 965–990.

[224] R.A. Maronna, R.D. Martin, and V. Yohai (2006). *Robust Statistics. Theory and Methods*. Wiley.

[225] T.A. Marques (2004). Predicting and correcting bias caused by measurement error in line transect sampling using multiplicative error model. *Biometrics* 60, 757–763.

[226] R.D. Martin and R.H. Zamar (1993). Efficiency-constrained bias-robust estimation of location. *Ann. Statist.* 21, 991–1017.

[227] J.T. Mayer (1750). Abhandlung über die Umwälzung des Monds um seine Axe. *Kosmographische Nachrichten und Sammlungen* 1, 52–183.

[228] I. Mizera and C.H. Mueller (1999). Breakdown points and variation exponents of robust M-estimators in linear models. *Ann. Statist.* 27, 1164–1177.

[229] I. Müller (1996). *Robust Methods in the Linear Calibration Model*. PhD Thesis, Charles University in Prague.

[230] J. Möttönen and H. Oja (1995). Multivariate spatial sign and rank methods. *J. Nonpar. Statist.* 5, 201–213.

[231] R. Navrátil (2012). Rank tests and R-estimates in location model with measurement errors. Financial Mathematics in Practice I. *Proceedings of Workshop of the Jaroslav Hájek Center*, pp. 37–44.

[232] R. Navrátil and A.K.Md.E. Saleh (2011). Rank tests of symmetry and R-estimation of location parameter under measurement errors. *Acta Univ. Palacki. Olomuc., Fac. rer. nat., Mathematica* 50(2), 95–102.

[233] G. Neuhaus and L.-X. Zhu (1999). Permutation tests for multivariate location problems. *J. Multiv. Anal.* 69, 167–192.

[234] T. Nummi and J. Möttönen (2004). Estimation and prediction for low degree polynomial models under measurement errors with an application to forest harvesters, *Appl. Statist.* 53, Part 3, 495–505.

[235] R.L. Obenchain (1971). Multivariate procedures invariant under linear transformations. *Ann. Math. Statist.* 42, 1569–1578.

[236] H. Oja (1987). On permutation tests in multiple regression and analysis of covariance problems. *Austr. J. Statist.* 29, 91–100.

[237] H. Oja (1983). Descriptive statistics for multivariate distributions. *Statist. Probab. Lett.* 1, 327–332.

[238] H. Oja (1999). Affine invariant multivariate sign and rank tests and corresponding estimates: A review. *Scandinavian Journal of Statistics* 26, 319–343.

[239] H. Oja (2010). Multivariate methods in nonparametrics, an approach based on spatial signs and ranks. *Lecture Notes in Statistics* 199, Springer.

[240] H. Oja and R.H. Randles (2004). Multivariate nonparametric tests. *Statist. Sci* 4, 598–605.

[241] H. Oja, J. Möttönen and J. Tienari (1997). On the efficiency of multivariate spatial sign and rank tests. *Ann. Statist.* 25, 542–552.

[242] E. Parzen (1975). Nonparametric statistical data modelling. *J. Amer. Statistical Assoc.* 74, 105–131.

[243] E.S. Pearson (1931). The analysis of variance in cases of nonnormal variation. *Biometrika* 23, 114–133.

[244] M.S. Pinsker (1960). *Information and Information Stability of Random Variables and Processes* (in Russian). Izv. Akad. Nauk, Moscow.

[245] S. Portnoy (1991). Asymptotic behavior of the number of regression quantile breakpoints. *J. Sci. Statist. Comput.* 12, 867–883.

[246] S. Portnoy and J. Jurečková (1987). Asymptotics for one-step M-estimators in regression with application to combining efficiency and high breakdown point. *Comm. Statist. A* 16, 2187–2199.

[247] S. Portnoy and J. Jurečková (1999). On extreme regression quantiles. *Extremes* 2:3, 227–243.

[248] M.L. Puri and P.K. Sen (1971). *Nonparametric Methods in Multivariate Analysis.* J. Wiley, New York.

[249] S.T. Rachev (1991). *Probability Metrics and the Stability of Stochastic Models.* John Wiley & Sons, Chichester, U.K.

[250] R.H. Randles and D. Peters (1990). Multivariate rank tests for the two-sample location problem. *Comm. Stat. Theory Methods* 19, 4225–4238.

[251] R Development Core Team (2004). *R: A language and environment for statistical computing.* R Foundation for Statistical Computing, Vienna, http://www.R-project.org.

[252] J.A. Reeds (1985). Asymptotic number of roots of Cauchy location likelihood equations. *Ann. Statist.* 13, 775–784.

[253] R.D. Reiss (1989). *Approximate Distributions of Order Statistics.* Springer, New York.

[254] H. Rieder (1994). *Robust Asymptotic Statistics.* Springer, New York.

[255] D.M. Rocke and S. Lorenzato (1995). A two-component model for measurement error in analytical chemistry. *Technometrics* 37/2, 176–184.

[256] E. Roelant, S. van Aelst (2007). An L1-type estimator of multivariate location and shape. *Statistical Methods and Applications* 15, 381–393.

[257] P.R. Rosenbaum (2005). An exact distribution free test comparing two multivariate distributions based on adjacency. *J. R. Statist. Soc. B* 67, 515–530.

[258] P.J. Rousseeuw (1982). Most robust M-estimators in the infinitesimal sense. *Zeitschrift f. Wahrscheinlichkeitsth. verw. Geb.* 61, 541–551.

[259] P.J. Rousseeuw (1984). Least median of squares regression. *J. Amer. Statist. Assoc.* 79, 871–880.

[260] P.J. Rousseeuw (1985). Multivariate estimation with high breakdown point. *Mathematical Statistics and Applications,* Vol. B (W. Grossmann et al., eds.), pp. 283–297. Reidel, Dordrecht.

[261] P.J. Rousseeuw and A.M. Leroy (1987). *Robust Regression and Outlier Detection.* John Wiley & Sons, New York.

[262] P.J. Rousseeuw and B.C. van Zomeren (1990). Unmasking multivariate outliers and leverage points. *J. Amer. Statist. Assoc.* 85, 633–639.

[263] P.J. Rousseeuw and K. van Driessen (1999). A fast algorithm for the minimum covariance determinant estimator. *Technometrics* 41, 212–223.

[264] P.J. Rousseeuw and V.J. Yohai (1984). Robust regression by means of *S*-estimators. *Robust and Nonlinear Time Series Analysis* (J. Franke, W. Härdle and R.D. Martin, eds.), pp. 256–272. Springer, New York.

[265] P. Royston (1982a). An extension of Shapiro and Wilk's *W* test for normality to large samples. *Appl. Statistics* 31, 115–124.

[266] P. Royston (1982b). Algorithm AS 181: The *W* test for normality. *Appl. Statistics* 31, 176–180.

[267] P. Royston (1995). A remark on algorithm AS 181: The *W* test for normality. *Appl. Statistics* 44, 547–551.

[268] D. Ruppert and R.J. Carroll (1980). Trimmed least squares estimation in the linear model. *J. Amer. Statist. Assoc.* 75, 828–838.

[269] M. Salibian-Barrera, G. Willems and R. Zamar (2008): The fast-τ estimator for regression. *J. of Computational and Graphical Statistics* 17, 659–682.

[270] A.K.Md.E. Saleh, J. Picek and J. Kalina (2012). R-estimation of the parameters of a multiple regression model with measurement errors. *Metrika* **75** , 311–328.

[271] D. Sarkar (2008). *Lattice: Multivariate Data Visualization with R.* Springer, New York.

[272] M.F. Schilling (1986). Multivariate two-sample tests based on nearest neighbors. *J. Amer. Statist. Ass.* 81, 799–806.

[273] P.K. Sen (1964). On some properties of the rank-weighted means. *J. Indian Soc. Agricul. Statist.* 16, 51–61.

[274] P.K. Sen (1978). An invariance principle for linear combinations of order statistics. *Zeit. Wahrscheinlich. Verw. Geb.* 42, 327–340.

[275] P.K. Sen (1980). On nonparametric sequential point estimation of location based on general rank order statistics. *Sankhyā* A 42, 201–218.

[276] P.K. Sen (1981). *Sequential Nonparametrics: Invariance Principles and Statistical Inference.* John Wiley & Sons, New York.

[277] P.K. Sen (1986). On the asymptotic distributional risk of shrinkage and preliminary test versions of maximum likelihood estimators. *Sankhyā* A 48, 354–371.

[278] P.K. Sen (1994). Isomorphism of quadratic norm and OC ordering of estimators admitting first-order representation. *Sankhyā* A 56, 465–475.

[279] P.K. Sen (2002). Shapiro-Wilk type goodness-of-fit tests for normality: Asymptotics revisited. *Goodness-of-Fit Tests and Model Validity* (C. Huber-Carol et al., eds.), pp. 73–88. Birkhäuser, Boston.

[280] P.K. Sen, J. Jurečková and J. Picek (2003). Goodness-of-fit test of Shapiro-Wilk type with nuisance regression and scale. *Austrian J. of Statist.* 32, No. 1 & 2, 163–177.

[281] P.K. Sen, J. Jurečková and J. Picek (2013). Rank tests for corrupted linear models. *Journal of the Indian Statistical Association* 51/1, 201–230.

[282] R.J. Serfling (1980). *Approximation Theorems of Mathematical Statistics.* John Wiley & Sons, New York.

[283] R.J. Serfling (2004). Nonparametric multivariate descriptive measures based on spatial quantiles. *J. Statist. Planning Infer.* 123, 259–278.

[284] R.J. Serfling (2010). Equivariance and invariance properties of multivariate quantile and related functions, and the role of standardization. *Journal of Nonparametric Statistics* 22, 915–936.

[285] S.S. Shapiro and M.B. Wilk (1965). An analysis of variance for normality (complete samples). *Biometrika* 52, 591–611.

[286] G.L. Shevlyakov and N.O. Vilchevski (2001). *Robustness in Data Analysis*. De Gruyter.

[287] G.R. Shorack (2000). *Probability for Statisticians*. Springer, New York.

[288] G.R. Shorack and J.A. Wellner (1986). *Empirical Processes with Applications to Statistics*. John Wiley & Sons, New York.

[289] A.F. Siegel (1982). Robust regression using repeated medians. *Biometrika* 69, 242–244.

[290] G.L. Sievers (1978). Estimation of location: A large deviation comparison. *Ann. Statist.* 6, 610–618.

[291] D.G. Simpson, D. Ruppert and R.J. Carroll (1992). On one-step *GM*-estimates and stability of inference in linear regression. *J. Amer. Statist. Assoc.* 87, 439–450.

[292] C.G. Small (1990). A survey of multidimensional medians. *Intern. Statist. Rev.* 58, 263–277.

[293] W.A. Stahel (1981). Breakdown of covariance estimators. Research Report 31, Fachgruppe für Statistik, ETH Zürich.

[294] R.J. Staudte and S.J. Sheather (1990). *Robust Estimation and Testing*. John Wiley & Sons, New York.

[295] C. Stein (1956). Inadmissibility of the usual estimator for the mean of multivariate distribution. *Proceedings of Third Berkeley Symposium on Mathematics Statistics and Probability,* Vol, 1, pp. 197–206. University of California Press, Berkeley, CA.

[296] C. Stein (1981). Estimation of the mean of a multivariate normal distribution. *Ann. Statist.* 9, 1135–1151.

[297] S.M. Stigler (1986). *The History of Statistics. The Measurement of Uncertainty before 1900*. Belknap Press of Harvard University Press, London.

[298] N. Sugiura (1965). An example of the two-sided Wilcoxon test which is not unbiased. *Ann. Inst. Statist. Math.* 17, 261–263.

[299] N. Sugiura, H. Murakami, S.K. Lee and Y. Maeda (2006). Biased and unbiased two-sided Wilcoxon tests for equal sample sizes. *Ann. Inst. Statist. Math.* 58, 93–100.

[300] M. Tableman (1994a). The influence functions for the least trimmed squares and the least trimmed absolute deviations estimators. *Statist. Probab. Lett.* 19, 329–337.

[301] M. Tableman (1994b). The asymptotics for the least trimmed absolute deviations (LTAD) estimator. *Statist. Probab. Lett.* 19, 387–398.

[302] L.D. Taylor (1973). Estimation by minimizing the sum of absolute errors. *Frontiers in Econometrics* (P. Zaemba, ed.), pp. 189–190. Academic Press, New York.

[303] A. Topchii, Y. Tyurin and H. Oja (2003). Inference based on the affine invariant multivariate Mann-Whitney-Wilcoxon statistic. *J. Nonparametr. Statist.* 14, 403–414.

[304] J.W. Tukey (1960). A survey of sampling from contaminated distribution. *Contributions to Probability and Statistics* (I. Olkin, ed.), Stanford University Press, Stanford, CA.

[305] J.W. Tukey (1975). Mathematics and the picturing of data. *Proc. Intern. Congress of Mathematicians* 2, Montréal: Canadian Mathematics Congress, pp. 523–531.

[306] J.W. Tukey (1977). *Exploratory Data Analysis.* Addison-Wesley, Reading, MA.

[307] J.W. Tukey and D.H. McLaughlin (1963). Less vulnerable confidence and significance procedures for location based on a single sample: Trimming/Winsorization I. *Sankhyā* A 25, 331–352.

[308] D.E. Tyler (1994). Finite sample breakdown points of projection based on multivariate location and scatter statistics. *Ann. Statist.* 22, 1024–1044.

[309] I. Vajda (1988). *Theory of Statistical Inference and Information.* Reidel, Dordrecht.

[310] A.W. van der Vaart and J.A. Wellner (1996). *Weak Convergence and Empirical Processes (With Applications to Statistics.)* Springer, New York.

[311] C. van Eeden (1972). Analogue, for signed rank statistics, of Jurečková's asymptotic linearity theorem for signed rank statistics. *Ann. Math. Statist.* 43, 791–802.

[312] Y. Vardi and C.H. Zhang (2000). The multivariate L_1-median and associated data depth. *Proceedings of National Academy of Sciences USA* 97, 1423–1426.

[313] W.N. Venables and B.D. Ripley (2002). *Modern Applied Statistics with S.* 4 th ed., Springer, New York.

[314] J.Á. Víšek (1995). On high breakdown point estimation. *Comp. Statist.* 11, 137–146.

[315] J.Á. Víšek (2000). On the diversity of estimates. *Comp. Statist. Data Anal.* 34, 67–89.

[316] J.Á. Víšek (2002a). The least weighted squares I. The asymptotic linearity of normal equations. *Bull. Czech Econometric Society* 9/15, 31–58.

[317] J.Á. Víšek (2002b). The least weighted squares II. Consistency and asymptotic normality. *Bull. Czech Econometric Society* 9/16, 1–28.

[318] B. von Bahr (1965). On the convergence of moments in the central limit theorem. *Ann. Math. Statist.* 36, 808–818.

[319] R. von Mises (1947). On the asymptotic distribution of differentiable statistical functions. *Ann. Math. Statist.* 35, 73–101.

[320] N.C. Weber (1980). A martingale approach to central limit theorems for exchangeable random variables. *Journal of Applied Probability* 17, 662–673.

[321] J.A. Wellner (1979). Permutation tests for directional data. *Ann. Statist.* 7, 929–943.

[322] A.H. Welsh (1986). Bahadur representation for robust scale estimators based on regression residuals. *Ann. Statist.* 14, 1246–1251.

[323] A.H. Welsh and E. Ronchetti (2002). A journey in single steps: robust one-step M-estimation in linear regression. *J. Statist. Planning Infer.* 103, 287–310.

[324] H. Wickham (2009). *ggplot2: Elegant Graphics for Data Analysis.* Springer-Verlag New York.

[325] D.P. Wiens and Z. Zheng (1986). Robust M-estimators of multivariate location and scatter in the presence of asymmetry. *Canad. J. Statist.* 14, 161–176.

[326] H. Witting and U. Müller-Funk (1995). *Mathematische Statistik II. Asymptotische Statistik: Parametrische Modelle und nichtparametrische Funktionale.* Teubner, Stuttgart, Germany.

[327] V.J. Yohai (1987). High breakdown point and high efficiency robust estimates for regression. *Ann. Statist.* 15, 642–656.

[328] V.J. Yohai and R.H. Zamar (1988). High breakdown-point estimates of regression by means of the minimization of an efficient scale. *J. Amer. Statist. Assoc.* 83/402, 406–413.

[329] R.H. Zamar (1989). Robust estimation in the errors–in–variables model. *Biometrika* 76, 149–160.

[330] B. Zhang, N. Cressie and D. Wunch (2018). Inference for errors-in-variables models in the presence of systematic errors with an application to a satellite remote sensing campaign. *Technometrics*, DOI: 10.1080/00401706.2018.1476268, in press.

[331] V.M. Zolotarev (1983). Probability metrics. *Theor. Probab. Appl.* 28, 278–302.

[332] Y. Zuo (2003). Projection depth functions and associated medians. *Ann. Statist.* 31, 1460–1490.

[333] Y. Zuo and X. He (2006). On the limiting distributions of multivariate depth-based rank sum statistics and related tests. *Ann. Statist.* 34, 2879–2896.

Author index

Subject index

asymptotic distribution, 206
Andrews sinus function, 50, 92

asymptotic distribution, 46, 48,
    173–175, 178–180, 193, 203, 207–209
asymptotic normal distribution,
    79, 208
asymptotic relations, 185
asymptotic relative efficiency, 179, 206,
    209
asymptotic representation, 173
asymptotic variance, 176, 178
asymptotically equivalent estimator,
    120, 182, 183

balanced design, 97
bisquare function, 50
BLUE, 78
breakdown point, 33, 34, 47–49, 68, 69,
    74, 76, 77, 85, 147
    finite sample, 191
    in regression, 109, 131, 132, 134, 143
    of M-estimator of location, 46

Cauchy likelihood, 50, 66, 92
characteristics of robustness, 25
    quantitative, 30, 33
contaminated distribution, 47, 49,
    184–186
contamination model, 183

datasets
    chem, 64
    CYGOB1, 92
    engel, 103, 112, 117, 120, 124, 127,
        131, 133, 135, 144, 216
    father.son, 137, 199, 217

SAT, 153
    stackloss, 139, 144
    starsCYG, 138, 199, 217
depth, 163, 166, 167
Dirac probability measure, 18
distance of measures, 12, 13, 16, 21
    Lévy, 13
    Prochorov, 13
    Hellinger, 13, 14
    Kolmogorov, 13, 20, 53
    Kullback-Leibler divergence, 14, 217
    Lipschitz, 14
    relations, 15
    total variation, 13
dual program, 125

empirical distribution function, 11, 20
empirical probability distribution, 11,
    12, 19–21
empirical quantile function, 72
estimators
    of real parameter, 43
estimators with high breakdown points,
    131
expected value, 47
exponential tails, 76, 96–98

finite sample minimax, 53
Fisher consistency, 12, 13, 44, 45, 62,
    175
Fisher information, 48, 120, 164, 176,
    178, 181, 184, 204, 205, 207, 216

geometric mean, 11
Gini mean difference, 40, 72, 87
global sensitivity, 30, 47–49, 76, 77
GM-estimator, 40, 111, 115–117
GR-estimator, 118